PLANT SCIENCE RESEARCH AND PRACTICES

RUBUS: AN OVERVIEW

Plant Science Research and Practices

Additional books and e-boooks in this series can be found on Nova's website under the Series tab.

PLANT SCIENCE RESEARCH AND PRACTICES

RUBUS: AN OVERVIEW

DAVET CHABOT
EDITOR

NOTICE TO THE READER

Library of Congress Cataloging-in-Publication Data

Names: Chabot, Davet, editor.
Title: Rubus: an overview / Davet Chabot.
Other titles: Plant science research and practices.
Description: Hauppauge, New York : Nova Science Publishers, [2020] | Series: Plant science research and practices | Includes bibliographical references and index. |
Identifiers: LCCN 2020001388 (print) | LCCN 2020001389 (ebook) | ISBN 9781536173765 (paperback) | ISBN 9781536173772 (adobe pdf)
Subjects: LCSH: Rubus. | Blackberries.
Classification: LCC QK495.R78 R83 2020 (print) | LCC QK495.R78 (ebook) | DDC 583/.642--dc23
LC record available at https://lccn.loc.gov/2020001388
LC ebook record available at https://lccn.loc.gov/2020001389

Published by Nova Science Publishers, Inc. † New York

CONTENTS

PREFACE

Rubus blackberries are grown worldwide and are known for their excellent nutritional and bioactive characteristics, particularly due to their high content of nutrients such as sugars, dietary fibre, minerals, and phenolic compounds. Rubus: An Overview summarizes the available data on nutrients, bioactive compounds, and health-promoting properties of the most studied Rubus blackberries.

Similarly, the authors review the role of the phenolic compounds in Rubus blackberries in the prevention and treatment of neurodegenerative disorders. Experimental studies suggest that anthocyanins of these fruits can improve cognition and human health.

The main pathogens and phytoparasites that cause damage in raspberries, thus causing important economic losses in the crop, are also received.

The concluding chapter suggests that increasing the role of genomics and genome editing technologies in Rubus presents new opportunities to develop more focused molecular tools for gene discovery and deployment.

Chapter 1 - *Rubus* blackberries are grown worldwide and are known for their excellent nutritional and bioactive characteristics, especially due to the high contents of nutrients such as sugars, dietary fibre, minerals, and phenolic compounds. The chemical composition of these berries has been related to biological effects, observed both *in vitro* and *in vivo*. Therefore, this chapter will summarize available data on nutrients, bioctive compounds, and health-promoting properties (antioxidant, anti-inflammatory, anticarcinogenic, antiatherosclerotic, and antimicrobial activities) of the most studied *Rubus* blackberries (*Rubus ulmifolius* Schott, *Rubus fruticosus* L., *Rubus adenotrichus* Schltdl., and *Rubus glaucus* Benth).

Chapter 2 - The bioactive compounds present in the genus *Rubus,* mainly, phenolic compounds such as flavonoids (anthocyanins, flavanols and phenolic acids) and ellagitannins, have garnered special attention due to the health benefits of their consumption. However, their biological activity is related with their absorption and metabolism. The purpose of this chapter is to summarize the available scientific information of the bioactive compounds in *Rubus* (blackberries and raspberries) and their bioavalibility, as well as their effect on intestinal microbiota and tissue distribution. Similarly, the authors review the role of the phenolic compounds in *Rubus* berries in the prevention and treatment of neurodegenerative disorders. Experimental studies suggest that anthocyanins of these fruits improve cognition and human health.

Chapter 3 - In marketable crops, fruit characteristics such as taste, size, and color must be adjusted to consumer demands. This positions the species from the genus *Rubus* with interesting perspectives due to its versatility to satisfy different markets. Species of this genus have been selected and backcrossed to obtain homogeneous crops with fruit that fits the demanded characteristics; however, after repeated cycles of selection and breeding, the species's genetic variability decreases and its susceptibility to pathogens increases, diminishing the crops' viability, vigor, and yield.

The genus *Rubus* includes several varieties differing in maturity, productivity, and with a variable tolerance to different diseases. Because of this, growers are constantly seeking for new varieties with better organoleptic qualities and high resistance to diseases. Unfortunately, a number of diseases associated to more than 740 species of bacteria, fungi, virus, and nematode agents have been described throughout the world, mainly affecting fruit quality in varieties of the genus *Rubus*. Those pathogens cause tissue necrosis, atrophy or deformation of organs, wilt, fruit crumbling and dwarfing, galls, root damage, and postharvest damage among other diseases.

This book chapter is aimed to review the main pathogens and phytoparasites that cause damage in raspberry (*Rubus idaeus*) causing important economic losses in the crop. The authors also present a synthesiezed view of the various etiological agents that affect the genus *Rubus*.

Chapter 4 - *Rubus* is the main member of Rosaceae family having basic chromosome number seven. It is native to temperate and subtropical regions of eastern Asia. *Rubus* is a medicinally important wild fruit crop used in the treatment of fever, cough, sore throat, epithelial cancer and cardiovascular disease. Important traits in *Rubus* are genetically encoded i.e., Yield, insect-pest resistance and quality linked parameters. But breeding in *Rubus* is tedious and time consuming due to highly heterozygous nature of this perennial fruit crop. Recent advances in the development of DNA or molecular markers have irreversibly changed the disciplines of breeding in *Rubus*. Increasing the role of genomics and genome editing technologies in *Rubus* presents new opportunities to develop more focused molecular tools for gene discovery and deployment which is not possible with conventional breeding techniques. So, the advancement in genomic information and modifications in linked properties directly improve crop by improving various traits in *Rubus*.

In: Rubus: An Overview
Editor: Davet Chabot

ISBN: 978-1-53617-376-5

Chapter 1

RUBUS BLACKBERRIES: AN OVERVIEW

Mayara Schulz*[*]*, Siluana Katia Tischer Seraglio, Luciano Valdemiro Gonzaga, Ana Carolina Oliveira Costa and Roseane Fett

Department of Food Science and Technology,
Federal University of Santa Catarina, Florianopolis, SC, Brazil

ABSTRACT

Rubus blackberries are grown worldwide and are known for their excellent nutritional and bioactive characteristics, especially due to the high contents of nutrients such as sugars, dietary fibre, minerals, and phenolic compounds. The chemical composition of these berries has been related to biological effects, observed both *in vitro* and *in vivo*. Therefore, this chapter will summarize available data on nutrients, bioctive compounds, and health-promoting properties (antioxidant, anti-inflammatory, anticarcinogenic, antiatherosclerotic, and antimicrobial activities) of the most studied *Rubus* blackberries (*Rubus ulmifolius*

[*] Corresponding Author's Email: schulzmay@gmail.com.

Schott, *Rubus fruticosus* L., *Rubus adenotrichus* Schltdl., and *Rubus glaucus* Benth).

Keywords: *Rubus ulmifolius*, *Rubus fruticosus*, *Rubus adenotrichus*, *Rubus glaucus*, berries

INTRODUCTION

The berries represent an important component of a healthy diet, and their consumption in sufficient amount reduces the risk of various diseases (Baby, Antony, and Vijayan 2018). In recent years, there has been growing interest in these fruits since contain not only crucial dietary components, but also a wide range of phytochemicals (Yeung et al. 2019; Baby, Antony, and Vijayan 2018).

The *Rubus* genus is one of the most diverse in morphological and genetic terms, which include various wild and cultivated species distributed globally. As a result, this genus presents a wide range of fruit types, as well as a significant variation in the nutrients and phytochemicals content (Garzón, Riedl, and Schwartz 2009). Blackberries are one of the most consumed berries of the *Rubus* genus (Azofeifa et al. 2015), and the best known and most studied species are *Rubus ulmifolius* Schott, *Rubus fruticosus* L., *Rubus adenotrichus* Schltdl., and *Rubus glaucus* Benth (Lee, Dossett, and Finn 2012; Schulz and Chim 2019).

Rubus ulmifolius Schott is native to Europe and North America and popularly known as elm-leaf blackberry. The fruits of this species are an aggregate of small drupelets, globoses, which during ripening turn from green to black (Reidel et al. 2016; Schulz and Chim 2019).

Rubus fruticosus L. blackberry, native to Europe, is formed by a dense cluster of separate units or drupelets that changes colour black or dark purple from red when the fruit is ripe. Fruiting occurs in both spring and autumn (Zia-Ul-Haq et al. 2014; Radovanović et al. 2013).

Rubus adenotrichus Schltdl. is native from Mexico to Ecuador with geographical distribution, especially at heights above 2000 meters above sea level. The fruits are small, polidrupas of colour black or red with seeds that present a hard and impermeable cover. The common names are mora, common mora, zarzamora, and mora vinera (Redondo et al. 2017; Schulz et al. 2019; Peraza-Padilla and Orozco-Aceves 2018).

Rubus glaucus Benth, commonly known as Andes berry or Castilla blackberry, is native to the Andes in northern South America where it grows year round, and present dark-red color when ripe. This fruit is also appreciated for its tartness, juiciness, and superior flavor in comparison to the other cultivated blackberries (Vergara, Vargas, and Acuña 2016; Garzón, Riedl, and Schwartz 2009).

Rubus blackberries have been introduced onto the local and world food stage and are usually sold for fresh consumption or as processed products specially such as jams, jellies, syrups, and wines (Schulz and Chim 2019). The tropical highland blackberries (*R. adenotrichus* and *R. glaucus*) species due higher acidity and a distinctive flavor are used mainly by juice industries for blends (Acosta-Montoya et al. 2010).

The health-promoting properties reported for *Rubus* blackberries are especially attributed to the high content of phenolic compounds, which are well known due to their high antioxidant capacity. These compounds can protect against the excess of free radicals and reactive species, which can help reduce the risk of disease development (Hidalgo and Almajano 2017). Also, *Rubus* blackberries contain other interesting chemical compounds such as sugars, minerals, dietary fibre, vitamin C, among others that can also contribute to human health (Schulz and Chim 2019).

Several studies have been conducted providing knowledge to support the high dietary and health values of *Rubus* blackberries. Thus, this chapter aimed to review the literature on chemical composition and antioxidant, anti-inflammatory, anticarcinogenic, antiatherosclerotic, and antimicrobial activities reported for *Rubus ulmifolius* Schott, *Rubus fruticosus* L., *Rubus adenotrichus* Schltdl., and *Rubus glaucus* Benth.

COMPOSITION

Water

Fruits have water as their main component, usually from 75 to 93 g 100 g^{-1} of edible portion (Hui et al. 2010). For *Rubus* blackberries, in 100 g, there are 70.8 – 89.4 g of water, with *R. ulmifolius* and *R. glaucus* having the highest content (Table 1).

Water has fundamental roles as a substrate or vehicle of biological reactions, maintenance of the cell and tissue integrity, as well as temperature and pH regulator in the human body (Hui et al. 2010).

Sugars

After water, the main component found in the fruits are carbohydrates, which are the main source of energy in the human diet and, in general, are classified into monosaccharides, oligosaccharides, and polysaccharides (Hui et al. 2010).

Glucose and fructose are the main monosaccharides, and their concentrations, as well as relative abundance, may change from one fruit to another (Milivojević et al. 2011). This occurs with *Rubus* blackberries since *R. ulmifolius* and *R. fruticosus* species showed higher fructose concentrations when compared to glucose, and the opposite occurred with *R. adenotrichus* and *R. glaucus* species (Table 1).

The sucrose, the most abundant oligosaccharide in fruits, was detected only in *R. ulmifolius* and *R. fruticosus* blackberries with values around 0.3 g 100 g^{-1} fresh weight (FW) (Table 1), which are lower than those found for most fruits, which are generally >1 g 100 g^{-1} FW (Hui et al. 2010).

Dietary Fiber

The consumption of dietary fibers in sufficient amounts can contribute to reducing glucose absorption, cholesterol and triglycerides levels, arterial tension, as well as prebiotic effect, which contributes to reducing the risk of chronic diseases (Zhang et al. 2018).

To date, to the best of our knowledge, dietary fiber content has been described only for *R. adenotrichus* and *R. glaucus* blackberries being the highest levels for the first species (6.3 to 6.5 g 100 g^{-1} FW). These values are higher than those described for other berries such as strawberry (1.3 g 100 g^{-1} FW), blueberry (1.9 g 100 g^{-1} FW), cherry (2.1 g 100 g^{-1} FW), and raspberry (5.8 g 100 g^{-1} FW) (de Souza et al. 2014).

Protein

Proteins have a wide structural variety since there is an expressive number of possibilities of amino acid sequences. Considering the wide structural variety, proteins perform various biological functions, including tissue structure, catalytic activity, efficient transport of molecules, and participation in the regulation of cellular or physiological activity (Moraes et al. 2013).

The study of proteins and amino acids in *Rubus* blackberries is still little explored since only crude protein data are still available. As shown in Table 1, the values ranged from 0.9 to 2.8 g 100 g^{-1} FW between species, with *R. ulmifolius* and *R. fruticosus* having the highest contents. Generally, fruits have values between 0.1 and 1 g 100 g^{-1} FW, while blackberries and raspberries showed values over 1 g 100 g^{-1} FW (Hui et al. 2010; de Souza et al. 2014).

Lipids

As for proteins, the highest total lipids content were found for blackberries *R. ulmifolius* and *R. fruticosus*, with values ranging from 1.2 to 1.9 g 100 g^{-1} FW (Table 1), which are superior to those reported for most fruits (usually <1 g 100 g^{-1} FW) (Hui et al. 2010; de Souza et al. 2014).

The fatty acids evaluation has become an important parameter since these compounds are crucial for the maintenance of normal physiological health. The consumption in adequate amount of mono and polyunsaturated fatty acids is associated with lower risk of development of diseases, especially cardiovascular diseases and atherosclerosis (Cheng, Wang, and Shao 2016; Ooi et al. 2015). For *Rubus* blackberries, fatty acids were evaluated in *R. ulmifolius*, which showed a predominance of polyunsatured (61.9 to 71.4% of the total lipid content), followed by monounsaturated (18.8 to 23.2% of the total lipid content) and saturated (9.7 to 14.9% of the total lipid content) fatty acids. The main fatty acids found for this species were linoleic acid (48.6 to 52.4%), oleic acid (18.4 to 22.6%), linolenic acid (13.3 to 18.6%), and palmitic acid (5.3 to 7.0%) (Da Silva et al. 2019; Morales et al. 2013).

Vitamin C

The most important physiological function of vitamin C is related to its high antioxidant activity, which contributes to the maintenance of the redox balance, protecting the cell membranes and lipoproteins against the lipid peroxidation (Figueroa-Méndez and Rivas-Arancibia 2015).

Ramful et al. (2011) classified fruits into three groups according to the vitamin C content: low level (<30 mg 100 g^{-1}), medium level (30 to 50 mg 100 g^{-1}), and high level (>50 mg 100 g^{-1}). According to this classification, except for *R. glaucus* that showed 93 mg 100 g^{-1} FW, *Rubus* blackberries qualify as fruits with low level of vitamin C since showed values <27 mg 100 g^{-1} FW (Table 1).

Minerals

Table 1. Water, sugars, dietary fibre, protein, lipids, vitamin C, and minerals of *Rubus* blackberries

	R. ulmifolius	*R. fruticosus*	*R. adenotrichus*	*R. glaucus*
Water (g 100 g^{-1} FW)	70.7 – 87.8	85.0 – 89.4	78.7 – 85.9	86.8 – 88.3
Fructose (g 100 g^{-1} FW)	3.2 – 7.8	7.6	1.5 – 4.6	1.8
Glucose (g 100 g^{-1} FW)	2.3 – 8.1	6.4	1.7 – 5.0	2.1
Sucrose (g 100 g^{-1} FW)	<LOD – 0.3	0.3	ND	ND
Dietary fiber (g 100 g^{-1} FW)	-	-	6.3 – 6.5	3.9
Protein (g 100 g^{-1} FW)	2.4 – 2.7	2.6	0.9 – 1.4	0.9
Lipids (g 100 g^{-1} FW)	1.2 – 1.9	1.8	0.1 – 1.0	0.1
Vitamin C (mg 100 g^{-1} FW)	17.1 – 26.8	7.1 – 10.1	20.0	92.9
Potassium (mg 100 g^{-1} DW)	860 – 1540	754 – 1016	-	720
Calcium (mg 100 g^{-1} DW)	433 – 620	154 – 207	-	371
Magnesium (mg 100 g^{-1} DW)	149 – 707	141 – 166	-	189
Sodium (mg 100 g^{-1} DW)	70 – 168	4.4 – 9.8	-	7.6
Zinc (mg 100 g^{-1} DW)	17.5	1.9 – 2.9	-	2.1
Iron (mg 100 g^{-1} DW)	2.1	3.8 – 8.4	-	16.7
Manganese (mg 100 g^{-1} DW)	<LOD – 1.4	1.3 – 3.1	-	2.4
References	(Ahmad et al. 2015; Ruiz-Rodriguez et al. 2014; Morales et al. 2013; Schulz et al. 2019; Da Silva et al. 2019)	(Van de Velde et al. 2016; Ogawa et al. 2008; Plessi, Bertelli, and Albasini 2007; Garzón, Riedl, and Schwartz 2009; Milivojević et al. 2011)	(Soto et al. 2019; Azofeifa et al. 2016; Acosta-Montoya et al. 2010)	(Leterm et al. 2006; Alarcon-Barrera et al. 2018; Soto et al. 2019)

FW: Fresh weight; DW: Dry weight; ND: Not detected.

An adequate consumptiom of minerals is essential for the high nutritional quality of the diet and also contributes to the prevention of diseases. These nutrients act in numerous organic processes, including hydroelectrolytic balance, blood pressure control, plus normal blood clotting, normal muscle, bone, and nervous system functioning, besides intervening in the cardiovascular and immune system and serve as constituents of prosthetic groups in metalloproteins and as activators of enzyme reactions (Akinyele and Shokunbi 2015; Costa et al. 2011).

For *R. adenotrichus*, to the best of our knowledge, mineral contents have not yet been published, while for *R. ulmifolius*, *R. fruticosus*, and *R. glaucus* blackberries the order of the levels of the main minerals was potassium > calcium > magnesium > sodium > zinc > iron > manganese (Table 1). Except for iron, which had higher concentrations in *R. glaucus* blackberry (16.7 mg 100 g^{-1} dry weight - DW), the highest values for all minerals were found for *R. ulmifolius* (Table 1).

In comparison to other fruits such as banana, apple, pineapple, grape, peach, pear, and plum the levels of calcium, magnesium, zinc, and iron were higher in *Rubus* blackberries, once the mentioned fruits show lower contents (<100 mg 100 g^{-1} DW for calcium and magnesium; <2 mg 100 g^{-1} DW for zinc and iron) (Hui et al. 2010). The levels of potassium, calcium, and magnesium in *Rubus* blackberries were also higher than berries such as red raspberry, strawberry, blueberry, and cherry, which showed 570.2 to 694.3, 0 to 62.1, and 40 to 140 mg 100 g^{-1} DW for potassium, calcium, and magnesium, respectively (de Souza et al. 2014).

Phenolic Compounds

Phenolic compounds are secondary metabolites involved in many functions in plants, including plant pigmentation, attraction of pollinators, antioxidant activity, protection against lipid peroxidation, ultraviolet radiation, among others (Taiz and Zeiger 2009; Pereira-Netto 2018). These compounds are the phytochemical components most investigated in fruits, mainly due to their antioxidant activity and potential action in the prevention of diseases and disorders (Dabbou et al. 2017; Pereira-Netto 2018). Among the fruits with high levels of phenolic compounds, *Rubus* blackberries can be highlighted as important sources of these components.

Total phenolic content (TPC) in *Rubus* blackberries often ranges from 200 to 600 mg gallic acid equivalent (GAE) 100 g^{-1} FW. In *R. fruticosus* and *R. ulmifolius*, were reported TPC ranging between 171 and 758 mg GAE 100 g^{-1} FW (Sariburun et al. 2010; Yang and Choi 2017; Betta et al. 2018; Yilmaz et al. 2009; Contessa et al. 2013; Ruiz-Rodriguez et al. 2014; Pinto et al. 2018; Milivojević et al. 2011; Plessi, Bertelli, and Albasini 2007; Schulz et al. 2019). These values were higher than those found for *R. glaucus* (266 to 294 mg GAE 100 g^{-1} FW) (Alarcón-Barrera et al. 2018; Garzón, Riedl, and Schwartz 2009). TPC of these blackberries was similar or higher than those reported for other berries such as açaí (454 mg GAE 100 g^{-1} FW), jabuticaba (222 to 440 mg GAE 100 g^{-1} FW), jambolan (146 to 185 mg GAE 100 g^{-1} FW), guabiju (292 mg GAE 100 g^{-1} FW), and blueberry (305 mg GAE 100 g^{-1} FW) (de Souza et al. 2014; Seraglio et al. 2018; Rufino et al. 2010). To the best of our knowledge, TPC for *R. adenotrichus* has not yet been published.

As shown in Table 2, phenolic compounds commonly found in *Rubus* blackberries belong to the classes of flavonoids (anthocyanin and non-anthocyanin compounds), phenolic acids, and ellagitannins.

Anthocyanins are an important group of phenolic compounds found in *Rubus* blackberries (Da Silva et al. 2019; Pinto et al. 2018). These glycosylated flavonoids are responsible for the shades of blue, red, and purple of the berries and are considerate markers of ripening since in many fruits, including in blackberries, their accumulation increase with the advance of ripening (Jaakola 2013; Chaves-Silva et al. 2018). Also, anthocyanins are compounds with potential health-benefits. Association between the consumption of anthocyanins and the reduction of incidences of cardiovascular disease has been reported (Fang 2015; Cassidy 2018).

In *Rubus* blackberries, total monomeric anthocyanins content (TMA) frequently ranges between 80 and 140 mg cyanidin-3-*O*-glucoside equivalents (Cy-3-gly) 100 g^{-1} FW. Higher values of TMA were found for *R. fruticosus* and *R. ulmifolius* (83 to 142 mg Cy-3-gly 100 g^{-1} FW) compared to those reported for *R. glaucus* and *R. adenotrichus* (45 to 169 mg Cy-3-gly 100 g^{-1} FW) (Ogawa et al. 2008; Plessi, Bertelli, and Albasini 2007; Sariburun et al. 2010; Yang and Choi 2017; Contessa et al. 2013; Schulz et al. 2019; Soto et al. 2019; Garzón, Riedl, and Schwartz 2009; Alarcón-Barrera et al. 2018). Similar values of TMA were found in other berries, such as jabuticaba (58 to 144 mg Cy-3-gly 100 g^{-1} FW), murta (143 mg Cy-3-gly 100 g^{-1} FW), puçá-preto (103 mg Cy-3-gly 100 g^{-1} FW), raspberry (16 to 78 mg Cy-3-gly 100 g^{-1} FW), and guabiju (41 mg Cy-3-gly 100 g^{-1} FW) (Rufino et al. 2010; Seraglio et al. 2018; Plessi, Bertelli, and Albasini 2007).

Pelargonidins and cyanidins are the main pigments found in bright red colored fruits, including in *Rubus* blackberries (Jaakola 2013; Chaves-Silva et al. 2018). According to Table 2, ten cyanidins glycosides and two pelargonidins glycosides were found in *Rubus* blackberries. Cyanidins glycosides can be highlighted as the major anthocyanins found in these fruits. Cyanidin-3-*O*-glucoside was the predominant anthocyanin found in *R. adenotrichus*, *R. fruticosus*, and *R. ulmifolius* (up to 1800.0 mg kg^{-1} FW). For *R. glaucus*, cyanidin-3-*O*-rutinoside was the main anthocyanin (630.0 mg kg^{-1} FW), followed by cyanidin-3-*O*-glucoside (508.0 mg kg^{-1} FW). The quantitative and qualitative composition of anthocyanins in the ripe fruit is influenced by environmental and genetic factors (Chaves-Silva et al. 2018; Jaakola 2013). Light expose and temperature are important factors that impact on the anthocyanins biosynthesis. High temperature can decrease the content of anthocyanins, while light exposure can increase their concentrations, especially in fruit skin (Jaakola 2013; Fang 2015). These aspects can be related to the differences in the profile and content of anthocyanins observed among the *Rubus* blackberries species.

A large number of non-anthocyanin flavonoids and phenolic acids were also reported in *Rubus* blackberries (Table 2). Rutin (up to 16.0 mg kg^{-1} FW), isoquercitrin (up to 23.1 mg kg^{-1} FW), kaempferol and its derivates (up to 29.0 mg kg^{-1} FW), epicatechin (up to 68.0 mg kg^{-1} FW), quercetin and its derivates (up to 101.0 mg kg^{-1} FW) can be pointed as the main non-anthocyanin flavonoids found in these berries. In relation to phenolic acids, the major compounds were gallic acid (up to 138.0 mg kg^{-1} FW), caffeic acid (up to 736.0 mg kg^{-1} FW), ferulic acid (up to 757.0 mg kg^{-1} FW), *p*-coumaric acid (up to 877.0 mg kg^{-1} FW), and ellagic acid (up to 3930.0 mg kg^{-1} FW). *R. fruticosus* contains a higher content of many non-anthocyanin flavonoids and phenolic acids than the others *Rubus* blackberries. But *R. glaucus* can be highlighted as the *Rubus* blackberry species with the highest content of ellagic acid.

Table 2. Content (mg kg^{-1} fresh weight) of phenolic compounds in *Rubus* blackberries

	R. ulmifolius	*R. fruticosus*	*R. adenotrichus*	*R. glaucus*
Flavonoids				
Anthocyanins				
Cyanidin-*O*-di-hexoside	DE	-	-	-
Cyanidin-3-*O*-dioxayl-glucoside	DE	NQ	-	-
Cyanidin-3-*O*-glucoside	194.0	1070.0 – 1800.0	680.0	508.0
Cyanidin-3-*O*-malonylglucoside	-	58.0 – 71.0	-	ND
Cyanidin-3-*O*-(6-*O*-malonyl)glucoside	-	56.0	40.0	-
Cyanidin-3-*O*-β-(6''-(3-hydroxy-3-methylglutaroyl)-glucoside)	-	187.0	-	-
Cyanidin-3-*O*-rutinoside	-	ND	ND	630.0
Cyanidin-3-*O*-sambubioside	-	-	-	NQ
Cyanidin-3-*O*-sophoroside	-	ND	-	-
Cyanidin-3-*O*-xyloside	-	8.0 – 81.0	-	-
Cyanidin-3-*O*-xylorutinoside	-	-	-	NQ
Pelargonidin-3-*O*-glucoside	-	-	-	2.0
Pelargonidin-3-*O*-rutinoside	42.3	-	-	NQ
Non-anthocyanins				

Table 2. (Continued)

	R. ulmifolius	*R. fruticosus*	*R. adenotrichus*	*R. glaucus*
Catechin	0.2 – 12.1	1.9 – 4.1	-	-
Crysin	0.01 – 6.1	-	-	-
Eriodictyol-*O*-hexoside	DE	-	-	-
Epicatechin	0.3 – 45.9	3.6 – 24.6	5.1	68.0
Isoquercitrin	2.3 – 23.1	-	-	-
Isorhamnetin	0.07	-	-	-
Kaempferol	0.1 – 0.4	ND – 25.0	2.1 – 7.3	4.0
Kaempferol-3-*O*-glucoside	-	29.0	-	-
Kaempferol-*O*-di-hexoside	DE	-	-	-
Luteolin	< 0.01	-	-	-
Myricetin	-	3.1	-	-
Naringenin	0.02 – 0.05	-	-	-
Pinobanksin	0.02 – 0.07	-	-	-
Pinocembrin	0.1	-	-	-
Quercetin	1.2 – 101.0	ND – 74.7	1.3 – 57.0	59.0
Quercetin-glucoside malonate	-	23.0	-	-
Quercetin-HMG-glucoside	DE	-	-	-
Quercetin-HMG-rhamnoside	DE	-	-	-
Quercetin-3-*O*-rutinoside	-	29.0	-	-
Quercetin-3-*O*-pentoside	-	100.0	-	-
Quercetin-(6'(-3-hydroxy-3-methylglautaroyl)-galatoside)	-	67.0	-	-
Rutin	11.6 – 15.4	16.0	-	-
Sinapaldehyde	0.05	-	-	-
Taxifolin-*O*-hexoside	DE	-	-	-
Phenolic acids				
Caffeic acid	0.09	0.3 – 736.0	-	-
Caffeic acid hexoside	DE	-	-	-
4-*O*-caffeoylquinic acid	DE	-	-	-
t-Caftaric acid	-	1.0	-	-
Chlorogenic acid	-	DE	-	-

Table 2. (Continued)

	R. ulmifolius	*R. fruticosus*	*R. adenotrichus*	*R. glaucus*
p-Coumaric acid	0.05	ND – 877.0	-	4.0
3,4-Dihydroxybenzoic acid	0.5 – 1.5	-	-	-
Ellagic acid	-	1.5 – 290.0	2.0 – 13.5	3930.0
Ellagic acid pentoside	DE	-	-	-
Ellagic acid glucuronide	DE	-		-
Ferulic acid	0.5	413.0 – 757.0	-	-
Gallic acid	0.8 – 0.9	138.0	0.5	49.0
p-Hydroxybenzoic acid	-	ND	-	-
Salicylic acid	0.2 – 0.8	-	-	-
Sinapic acid	0.3	-	-	-
Syringic acid	0.04	3.7	-	-
Ellagitannins				
di-HHDP glucose	-	14.0	-	-
Lambertianin C	-	38.0 – 71.0	598.0	5200.0
Sanguiin H-6	-	123.0	420.0	2450.0 [a]
Sanguiin H-10	DE	-	-	-
References	(Van de Velde et al. 2019; Betta et al. 2018; Ruiz-Rodriguez et al. 2014; Schulz et al. 2019; Da Silva et al. 2019)	(Jakobek et al. 2009; Türkben et al. 2010; Pinto et al. 2018; Milivojević et al. 2011; Ogawa et al. 2008; Radovanović et al. 2013; Van de Velde et al. 2016; Yang and Choi 2017)	(Gancel et al. 2011; Mertz et al. 2007; Acosta-Montoya et al. 2010)	(Vasco et al. 2009; Garzón, Riedl, and Schwartz 2009; Mertz et al. 2007)

ND: Not detected; NQ: Not quantified; DE: Data of extracts; [a] Dry matter.

Another important group of phenolic compounds is ellagitannins. In *Rubus* blackberries, four ellagitannins were determined (Table 2). Lambertianin C and sanguiin H-6 were the main compounds reported in *R. adenotrichus*, *R. fruticosus*, and *R. glaucus*. In *R. ulmifolius*, only sanguiin H-10 was found. A high concentration of lambertianin C in *R. glaucus* and *R. adenotrichus* can be highlighted, indicating that this compound is one of the main phenolic compounds present in these berries. The presence of a high concentration of ellagitannins in these fruits is an important finding since strong *in vitro* and *in vivo* antioxidant effects have been related to these compounds (Landete 2011; Larrosa et al. 2010).

Cultivar, species, ripening stage, and edaphoclimatic conditions are some of the factors that can influence the content and the profile of phenolic compounds in fruits (Pereira-Netto 2018; Lee, Dossett, and Finn 2012; Taiz and Zeiger 2009). The combined action of these factors can directly result in the differences observed in the phenolic content among the *Rubus* blackberry species and into the same species.

HEALTH-PROMOTING PROPERTIES

Antioxidant

Compounds with antioxidant potential can inhibit oxidation processes and prevent chronic diseases related to oxidative stress, such as certain cancer types, atherosclerosis, and cardiovascular disease. These components can act as free radical quencher, singlet oxygen scavengers, or chelators of metal ions (Cömert and Gökmen 2018; Shahidi and Ambigaipalan 2015). Among the bioactive phytochemicals, the phenolic compounds can be highlighted due to their high correlations with the antioxidant activity (Yang and Choi 2017; Pereira-Netto 2018). The antioxidant activity of these compounds is basically related to the present of an aromatic ring with one or more hydroxyl groups that can stabilize and delocalize unpaired electrons (Hidalgo and Almajano 2017).

Rubus blackberries have been proving to be good sources of bioactive compounds, especially of phenolic compounds, and have been shown high antioxidant potential. In Table 3, the published studies on the antioxidant activity as well as others possible health-promoting properties are presented.

There are numerous methods for measuring *in vitro* antioxidant activity, which vary according to experimental conditions and reaction mechanisms (Ma et al. 2011; Shahidi and Ambigaipalan 2015). In this sense, it is recommended employ at least two assays to assess the antioxidant activity of samples (Kamiloglu et al. 2015). In *Rubus* blackberries, the assays commonly employed to evaluate this activity are based in the ferric reducing antioxidant power (FRAP) and in the deactivation of radicals such as 2,2-diphenyl-1-picrylhydrazyl (DPPH) and 2,2-azino-bis(3-ethylbenzothiazoline-6-sulphonic acid) (ABTS).

Rubus blackberries presented DPPH scavenging activity similar to each other. For *R. ulmifolius*, DPPH values between 26.3 and 159.1 mmol Trolox equivalent kg^{-1} FW were found (Ruiz-Rodriguez et al. 2014; Betta et al. 2018). These values were similar to those reported for *R. glaucus* when DPPH values between 51.7 to values close to 150 mmol Trolox equivalent kg^{-1} FW were reported (Horvitz, Chanaguano, and Arozarena 2017; Alarcón-Barrera et al. 2018). A wide range of DPPH scavenging activity was also found for *R. fruticosus*, ranging from 9 to 177.1 mmol Trolox equivalent kg^{-1} FW (Sariburun et al. 2010; Yang and Choi 2017). The evaluation of *R. adenotrichus* juice resulted in lower DPPH values (3.3 to 83.3 mmol Trolox equivalent kg^{-1} FW) (Gancel et al. 2011) compared to the other *Rubus* blackberries. But these values can be considerate satisfactory, suggesting higher DPPH scavenging activity in the whole *R. adenotrichus* fruit. The DPPH scavenging activity of *Rubus* blackberries was similar to other fruits such as yellow guava (87.7 to 100.0 mmol Trolox equivalent kg^{-1} FW), guabiju (142.9 to 167.8 mmol Trolox equivalent kg^{-1} FW), jambolan (107.3 to 143.8 mmol Trolox equivalent kg^{-1} FW), and raspberry (64.1 to 127.0 mmol Trolox equivalent kg^{-1} FW) (Sariburun et al. 2010; Betta et al. 2018).

Also, the methanolic extract of *R. ulmifolius* showed high DPPH scavenging activity (98.9% of inhibition at 400 $\mu g\ mL^{-1}$) (Ahmad et al. 2015) compared to anthocyanins extract of *R. fruticosus* (51% of inhibition at 2000 $\mu g\ mL^{-1}$) (Ogawa et al. 2008). These results suggest the possible contribution of other compounds, besides anthocyanins, in the scavenging activity of *Rubus* blackberries.

The ability in the deactivation of ABTS radical was also investigated in *Rubus* blackberries. In *R. ulmifolius*, ABTS values ranged from 22.8 and 88.9 mmol Trolox equivalent kg^{-1} FW (Ruiz-Rodriguez et al. 2014a). These values were higher than those reported for *R. glaucus* (20.1 mmol Trolox equivalents kg^{-1} FW) (Garzón, Riedl, and Schwartz 2009) but lower than those found for *R. fruticosus* (1.7 to 146.9 mmol Trolox equivalent kg^{-1} FW) (Pinto et al. 2018; Yang and Choi 2017; Sariburun et al. 2010). The ABTS scavenging activity of *Rubus* blackberries was in many cases higher than those reported for other fruits such as açaí (15.1 mmol Trolox equivalent kg^{-1} FW), acerola (96.6 mmol Trolox equivalent kg^{-1} FW), juçara (78.3 mmol Trolox equivalent kg^{-1} FW), murta (49.1 mmol Trolox equivalent kg^{-1} FW), jambolan (29.7 mmol Trolox equivalent kg^{-1} FW), jaboticaba (37.5 mmol Trolox equivalent kg^{-1} FW), and raspberry (64.3 to 117 mmol Trolox equivalent kg^{-1} FW) (Rufino et al. 2010; Sariburun et al. 2010).

About the reducing potential of the *Rubus* blackberries, high ferric reducing potential was reported for *R. ulmifolius* and *R. glaucus*. FRAP values ranging from 7.32 to 141.6 mmol Trolox equivalent kg^{-1} FW were found for *R. ulmifolius* (Betta et al. 2018; Ruiz-Rodriguez et al. 2014), while in *R. glaucus*, FRAP values ranged from 45.0 to values closed to 100 mmol Trolox equivalent kg^{-1} FW (Alarcón-Barrera et al. 2018; Garzón, Riedl, and Schwartz 2009). These values were higher than those found for *R. fruticosus*, which ranged from around 9 to 20 mmol Trolox equivalent kg^{-1} FW (Kostecka-Gugała et al. 2015; Yang and Choi 2017). In general, the ferric reducing potential of *Rubus* blackberries was higher than those values reported for other fruits such as yellow guava (0.9 to 1.9 mmol Trolox equivalent kg^{-1} FW), guabiju (5.4 to 7.3 mmol Trolox equivalent kg^{-1} FW), jambolan (2.6 to 3.4 mmol Trolox equivalent kg^{-1} FW), jabuticaba (8.2 mmol Trolox equivalent kg^{-1} FW), and acerola (40.7 to 59.1 mmol Trolox equivalent kg^{-1} FW) (Betta et al. 2018).

R. ulmifolius and *R. fruticosus* usually present better results compared to the other *Rubus* blackberries. Antioxidant activity of *Rubus* blackberries is commonly related to the profile and content of phenolic compounds, especially the anthocyanins. Usually, the highest antioxidant activity is observed in the blackberries with a higher content of polyphenols.

In vivo, the antioxidant potential of *Rubus* blackberries was also investigated in few studies. The *in vivo* antioxidant activity of *R. adenotrichus* beverage (RAB) was evaluated in a rat model by Azofeifa et al. (2016). Rats with streptozotocin-induced diabetes were treated orally for

40 days with RAB at 12.5 and 25% (v/v). Improvement of the plasma antioxidant capacity and the reducing of the lipid peroxidation in the kidney (19%) and the plasma (7.5%) were observed with the consumption of the higher dose of RAB (25%).

Hassimotto et al. (2008) investigated the *in vivo* antioxidant activity of *R. fruticosus* beverage in healthy human subjects. The beverages were prepared with *R. fruticosus* (corresponding to 400 mg of cyanidin equivalent 50 kg^{-1} of body weight) homogenized with water (200 mL) or defatted milk (200 mL). The intake of both beverages, with an interval of 4 weeks each one, resulted in an increase in the ascorbic acid and catalase activity in the plasma. Therefore, the relevance of other antioxidants beyond the phenolic compounds was suggested in this study, since a good correlation between the antioxidant activity and the ascorbic acid was observed in the human plasma.

The antioxidant potential of *Rubus* blackberries found in *in vitro* assays was also observed in *in vivo* studies. Despite the few *in vivo* studies, they indicate promising antioxidant potential of these berries and the influence of antioxidant components, such as anthocyanins, phenolic acids, flavonoids, ascorbic acid, and vitamin E, in the antioxidant activity of *Rubus* blackberries.

Anti-Inflamatory

Inflammation is a complex defensive response of the immunological system typically induced by microbial infections, allergens, tissue injury or trauma, among others. Inflammation involves the recruitment of immune cells, such as macrophages, by specific signals induced by the inflammatory factors. These cells secrete pro-inflammatory cytokines and chemokines, which attract the lymphocytes and trigger an immune response (Joseph, Edirisinghe, and Burton-Freeman 2016; Maleki, Crespo, and Cabanillas 2019).

When an overproduction of inflammatory mediators occurs, a dysregulated inflammatory response can be stimulated, resulting in the increasing of chronic inflammatory disorders. The chronic inflammation is characterized by high levels of pro-inflammatory mediators such as interferon, transforming growth factor-β, interleukins (IL), C-reactive protein, tumor necrosis factor-α, as well as nitric oxide (NO) and reactive oxygen species (ROS) (Zhang, Virgous, and Si 2019; Maleki, Crespo, and Cabanillas 2019). Phytochemicals, such as phenolic compounds, terpenes, protein inhibitors, organic acids, among others, are potential anti-inflammatory components since they can promote the inhibition of pro-inflammatory mediators (Zhang, Virgous, and Si 2019; Joseph, Edirisinghe, and Burton-Freeman 2016).

Despite the elevate levels of bioactive compounds in *Rubus* blackberries, few studies are still found related to the anti-inflammatory potential of these fruits and/or their extracts (Table 3). However, promising results have been found.

The potential influence of *R. adenotrichus* phenols on inflammatory processes was reported by Azofeifa et al. (2013). Extracts (acetone:water:formic acid, 70:30:2, v/v/v) of *R. adenotrichus* (50 and 100 μg mL^{-1}) inhibited IL-6 production in murine macrophage cell line J774A.1 and also showed inhibitory properties against superoxides and NO.

Ellagitannins are a group of phenolic compounds with biological activity found in *Rubus* blackberries. The anti-inflammatory activity of ellagitannin extracts was investigated in a rat model with ethanol-induced gastric lesions. Protective effect against ethanol injury was observed in rats treated orally for ten days with 20 mg kg^{-1} of ellagitannin extracts of *R. fruticosus*, reducing gastric lesions in 88% (Sangiovanni et al. 2013).

A high level of cyanidin-3-*O*-glucoside was also suggested as a possible compound responsible for the anti-inflammatory effects reported for *R. fruticosus*. Extracts (methanol:water, 80:20, v/v) of *R. fruticosus* (50 μg mL^{-1}) decreased NO production and the gene expression of Cox-2 (cyclooxygenase-2) and IL-6 in murine macrophage cell line RAW 264.7 (Van de Velde et al. 2016).

In another study, the ability of extracts (methanol:water:acetic acid, 80:20:0.5, v/v/v) of *R. fruticosus* (50 μg mL^{-1}) to inhibit the intracellular ROS production and NO synthesis was investigated *in vitro* in murine RAW

264.7 macrophage model. The crude extract and the polyphenol fractions (anthocyanin-enriched and proanthocyanidin-enriched fractions) decreased the ROS production (up to 50%). Due to the highest ROS reduction promoted by proanthocyanidin-enriched fraction, it was suggested that proanthocyanidin components might be more active than anthocyanins against the oxidative stress. *R. fruticosus* extracts also exhibited NO synthesis inhibition up to 25% and decreased the gene expression of Cox-2, iNOS (nitric oxide synthase), IL-6, and IL-1β (Van de Velde et al. 2019).

The *in vivo* anti-inflammatory activity of *R. fruticosus* was evaluated in a rat model. The animals were treated orally for nine days with 3 mL of *R. fruticosus* juice (RFJ). The results showed that RFJ inhibits the carrageenan-induced paw oedema in rats, from the first (64.3%) to the fifth (70.8%) hour (Monforte et al. 2018).

Considering the effects reported above, it is possible to assume that the profile and content of phenolic compounds of *Rubus* blackberries play key roles in the inflammatory processes.

Anticarcinogenic

Recently, there has been increasing interest in exploring natural sources like fruits that may have potential as either adjuvant or as chemopreventive agents to reverse, inhibit, or prevent the progression of cancer (Goh et al. 2019; Nowak et al. 2017).

Among the compounds present in fruits, polyphenols are the most studied concerning chemopreventive effects. The mechanisms of action of these compounds for chemoprevention include inhibition of oxidation, induction of antioxidant enzymes, induction of cell apoptosis, antiproliferation, anti-inflammatory effect, regulation of the immune system, and estrogenic/antiestrogenic activity (Ljevar et al. 2016).

In this context, due to the promising chemopreventive effects of polyphenols of fruits and their products, cytotoxicity effects of six selected fruit wines, including wine of blackberry (*R. fruticosus*), towards human breast (MCF-7), colon (CaCo-2) and cervical (HeLa) cancer cell lines was evaluated by Ljevar et al. (Ljevar et al. 2016) (Table 3). A different volume ratio of fruit wine (1 – 20%) was tested, and *R. fruticosus* wine inhibited the growth of all human cancer cells in a dose-dependent manner. Blackberry wine, in the volume ratio of 10 and 20%, showed to be the most effective, and HeLa and MCF-7 cells were more susceptible than CaCo-2 cells. These results were similar to those found for cherry, raspberry, and blackcurrant wines and can be associated with their contents of phenolic compounds. These four wine fruits showed a higher concentration of total phenolics and anthocyanins, as well as higher antioxidant capacity than strawberry and apple wines, which were less effective in the cytotoxicity assays.

Cytotoxic properties were also studied by Zambrano et al. (2018), which evaluated the effects of *R. glaucus* blackberry pasteurized pulp on the human breast (MCF-7 and SKBR3) and prostate (PC3) cancer cell lines (Table 3). Specifically at a concentration of 5% of the pulp, cell viability percentage for MCF-7 was 9.1%, while for SKBr3 was 8.3%. For PC3, cell line viability percentage was 11.7% at the same concentration (5% of the pulp). This study also evaluated *Annona muricata* L. (soursop) and found that the blackberry was more effective in inhibiting the growth of human breast cancer cell lines, with four times lower cell viability percentages.

Antiatherosclerotic

Dyslipidemia is characterized by hypertriglyceridemia, hepatic oversecretion of (apo) B100-containing lipoproteins, high levels of LDL-cholesterol, and low levels of HDL-cholesterol. These changes in blood lipids are considered primary risk factors for the development of cardiovascular diseases, including atherosclerosis. Also, another key factor in the initiation of atherosclerosis process is the oxidation of LDL within the vessel wall (Mulvihill and Huff 2010; Rahman, Murphy, and Woollard 2017).

Increased dietary intake of fruit phenolics has been linked to antiatherosclerotic effects, which are highly effective antioxidants capable of inhibiting LDL oxidation, as well as associated with decreased levels of blood lipids (Borochov-Neori et al. 2015). In this context, the hypolipidaemic effect of the consumption of *R. adenotrichus* blackberry was studied by Azofeifa et al. (Azofeifa et al. 2016) (Table 3). This study demonstrated that rats with streptozotocin-induced diabetes that given orally *R. adenotrichus* beverage (25%) for 40 days significantly decreased triglycerides (-43.5%), cholesterol (-28.6%), and lipid peroxidation (-19%) in plasma.

Antimicrobial

Despite the permitted use of synthetic antimicrobials as food preservatives, studies have been performed to evaluate the antimicrobial activities of natural sources, including fruits, for possible application in foods (Beristain-Bauza et al. 2019). The studies suggest that the antimicrobial potential of berries is mainly related to high polyphenol concentration and antioxidant activity (Das et al. 2017; Schulz and Chim 2019).

For *Rubus* blackberries, antimicrobial activity was studied for *R. ulmifolius* and *R. fruticosus* species. Hajaji et al. (2017) demonstrated that the *Rubus ulmifolius* methanolic extract had a significant antimicrobial activity against all the six pathogens studied (*Staphylococus aureus*, *Enterococcus feacium*, *Streptococcus agalactiae*, *Escherichia coli*, *Salmonella typhimurium*, and *Candida albicans*) (Table 3), especially *E. coli*, *S. agalactiae*, and *C. albicans*, which showed inhibition zone diameters (15 μl/disc) of 28, 50, and 39 mm, respectively.

Rubus fruticosus methanolic extract also showed significant antimicrobial activity (Table 3). Riaz et al. (2011) demonstrated that this extract was effective against *Streptococcus aureus*, *Micrococcus luteus*, *Citrobacter*, *Bacillus subtilis,* and *Pseudomonas aeruginosa* with inhibition zone diameters (20 μg/disc) of 8 to 8.1 mm. In contrast, there was no significant antimicrobial activity for *Salmonella typhi*, *E. coli*, and *Proteus mirabilis*, as well as antifungal activity for *Aspergilus parasiticus*, *Aspergilus niger*, *Yersinia aldovae*, *Candida albicans*, *Aspergillus effusus*,

Fusarium solani, *Macrophomina phaseolina*, *Saccharomyces cerevisiae*, and *Trichophyton rubrum*. Radovanović et al. (2013) observed that *Rubus fruticosus* formic acid:methanol:water (0.1:70:29.9, v/v/v) extract had high antimicrobial activity on *Clostridium perfringens*, *Bacillus subtilis*, *Listeria innocua*, *S. aureus*, *Sarcina lutea*, *Micrococcus flavus*, *E. coli*, *Pseudomonas aeruginosa*, *Salmonella enteritidis*, and *Shigella sonnei*. Inhibition zone diameters (50 µl/disc) ranged from 12.0 to 16.2 mm.

Table 3. Health-promoting properties tested with *Rubus* blackberries

Species	Activity	Positive effects	References
R. ulmifolius	Antioxidant	DPPH scavenging activity	(Ahmad et al. 2015)
		Reducing potential by FRAP assay	(Contessa et al. 2013)
		Reducing potential by FRAP assay; DPPH scavenging activity	(Betta et al. 2018)
		Reducing potential by FRAP and Folin-Ciocalteu assays	(Schulz et al. 2019)
		Reducing potential by FRAP and Folin-Ciocalteu assays; ABTS and DPPH scavenging activities	(Ruiz-Rodríguez et al. 2014)
		Inhibition of lipid peroxidation by β-carotene bleaching inhibition and TBARS assays	(Morales et al. 2013)
		ROS scavenging activity by hydroxyl radical, superoxide anion, and hydrogen peroxide assays	(Hajaji et al. 2017)
	Antimicrobial	Activity against *Staphylococus aureus*, *Streptococcus agalactiae*, *Escherichia coli*, *Enterococcus feacium*, *Salmonella typhimurium*, and *Candida albicans*	(Hajaji et al. 2017)

Species	Activity	Positive effects	References
R. fruticosus	Antioxidant	DPPH and oxygen (by ORAC assay) scavenging activities	(Hassimotto, Pinto, and Lajolo 2008)
		DPPH scavenging activity	(Jakobek et al. 2009)
		Reducing potential by FRAP assay	(Koczka, Stefanovits-Bányai, and Prokaj 2018)
		Reducing potential by FRAP and CUPRAC assays; DPPH scavenging activity	(Kostecka-Gugała et al. 2015)
		ABTS scavenging activity	(Milivojević et al. 2011)
		DPPH, ABTS, and oxygen (by ORAC assay) scavenging activities; inhibition of lipid peroxidation by β-carotene bleaching inhibition assay; reducing potential by FRAP assay	(Monforte et al. 2018)
		DPPH and ABTS scavenging activities; reducing potential by FRAP assay	(Ogawa et al. 2008)
		ABTS scavenging activity	(Pinto et al. 2018)
		DPPH scavenging activity	(Radovanović et al. 2013)
		Oxygen (by ORAC assay) scavenging activity	(Sangiovanni et al. 2013)
		DPPH and ABTS scavenging activities; reducing potential by CUPRAC assay	(Sariburun et al. 2010)
		DPPH scavenging activity; reducing potential by FRAP assay	(Van de Velde et al. 2016)
		DPPH and ABTS scavenging activities; reducing potential by FRAP assay	(Yang and Choi 2017)
		DPPH scavenging activity; inhibition of lipid peroxidation by β-carotene bleaching inhibition assay	(Yilmaz et al. 2009)

Table 3. (Continued)

Species	Activity	Positive effects	References
	Anti-inflamatory	Inhibited carrageenan-induced paw oedema in rats	(Monforte et al. 2018)
		Protective effect against ethanol injury in rats	(Sangiovanni et al. 2013)
		Decreased NO, Cox-2, and IL-6 in macrophage cell line RAW 264.7	(Van de Velde et al. 2016)
		Decreased ROS, NO, Cox-2, iNOS, IL-6, and IL-1β in macrophage cell line RAW 264.7	(Van de Velde et al. 2019)
	Anticarcinogenic	Inhibited proliferation of human cancer epithelial cell lines (MCF-7, CaCo-2, and HeLa)	(Ljevar et al. 2016)
	Antimicrobial	Antibacterial activity against *Streptococcus aureus*, *Micrococcus luteus*, *Citrobacter*, *Bacillus subtilis* and *Pseudomonas aeruginosa*	(Riaz, Ahmad, and Rahman 2011)
		Antibacterial activity against *Escherichia coli*, *Pseudomonas aeruginosa*, *Salmonella enteritidis*, *Salmonella enteritidis*, *Clostridium perfringens*, *Bacillus subtillis*, *Staphylococcus aureus*, *Listeria inocua*, *Sarcina lutea*, and *Sarcina lutea*	(Radovanović et al. 2013)
R. adenotrichus	Antioxidant	DPPH and ORAC scavenging activities	(Gancel et al. 2011)
		DPPH and oxygen (by ORAC assay) scavenging activities; inhibition of lipid peroxidation by TBARS assay; stability of catalase activity	(Azofeifa et al. 2016)

Species	Activity	Positive effects	References
		DPPH, NO, and oxygen (by ORAC assay) scavenging activities; inhibition of lipid peroxidation by TBARS assay; ROS scavenging activity by superoxide anion assay	(Azofeifa et al. 2013)
		Oxygen (by ORAC assay) scavenging activity	(Acosta-Montoya et al. 2010)
	Anti-inflamatory	Decreased superoxides, NO, and IL-6 in macrophage cell line J774A.1	(Azofeifa et al. 2013)
	Antiatherosclerotic	Decreased triacylglycerols and cholesterol in rats	(Azofeifa et al. 2016)
R. glaucus	Antioxidant	DPPH scavenging activity	(Horvitz, Chanaguano, and Arozarena 2017)
		Reducing potential by FRAP assay; ABTS scavenging activity	(Garzón, Riedl, and Schwartz 2009)
		Reducing potential by FRAP assay; DPPH scavenging activity; ROS scavenging activity by superoxide anion and hydrogen peroxide assay	(Alarcón-Barrera et al. 2018)
	Anticarcinogenic	Cytotoxicity for breast (MCF-7 and SKBR3) and prostate cancer (PC3) cell lines	(Zambrano et al. 2018)

FRAP: Ferric reducing antioxidant power; DPPH: 2,2-diphenyl-1-picrylhydrazyl; ORAC: Oxygen radical absorbance capacity; NO: Nitric oxide; ABTS: 2,2-azino-bis(3-ethylbenzothiazoline-6-sulphonic acid); ROS: Reactive oxygen species; CUPRAC: Cupric reducing antioxidant power; IL: Interleukins; TBARS: Thiobarbituric acid reactive substances; Cox-2: Cyclooxygenase-2; iNOS: Inducible nitric oxide synthase.

CONCLUSION

In this chapter, the chemical composition and biological potential of *Rubus* blackberries were summarized. In these berries were found a large variety of components, such as sugars, proteins, minerals, ascorbic acid, anthocyanins, non-anthocyanin flavonoids, phenolic acids, among others. In general, *R. fruticosus* and *R. ulmifolius* blackberries showed the highest levels of nutrients and phenolic compounds.

Different studies indicate promising *in vitro* and *in vivo* health-promoting properties properties of *Rubus* blackberries, such as antioxidant, anti-inflammatory, antimicrobial, and positive effects on blood lipids and atherosclerosis. It is important to note that the results published so far on biological activities for *Rubus* blackberries were interesting, but represent a very preliminary step in establishing the effectiveness of these fruits in human health. Studies performed with *in vitro* assays or cell lines are not equivalent to a clinical trial, in which humans consume *Rubus* berries or their products. *In vivo* studies are required to consider human digestion and absorption, and if *Rubus* bioactive compounds will be able to reach the sites of interest at a required concentration.

Therefore, these data suggest that *Rubus* blackberries are important sources of nutrients to the diet and contribute to potential human health-benefits. These findings reinforce the importance of further research, especially *in vivo*, for a better understanding of the beneficial properties of these berries, stimulating their *in natura* consumption and exploration as potential food ingredients.

References

Acosta-Montoya, Óscar, Fabrice Vaillant, Sonia Cozzano, Christian Mertz, Ana M. Pérez, and Marco V. Castro. 2010. "Phenolic content and antioxidant capacity of tropical highland blackberry (*Rubus adenotrichus* Schltdl.) during three edible maturity stages." *Food Chemistry* 119 (4): 1497–1501. doi: 10.1016/j.foodchem.2009.09.032.

Ahmad, Mushtaq, Saima Masood, Shazia Sultana, Taibi Ben Hadda, Ammar Bader, and Muhammad Zafar. 2015. "Antioxidant and Nutraceutical Value of Wild Medicinal *Rubus* Berries." *Pakistan Journal of Pharmaceutical Sciences* 28 (1): 241–47.

Akinyele, I. O., and O. S. Shokunbi. 2015. "Concentrations of Mn, Fe, Cu, Zn, Cr, Cd, Pb, Ni in Selected Nigerian Tubers, Legumes and Cereals and Estimates of the Adult Daily Intakes." *Food Chemistry* 173: 702–8. doi: 10.1016/J.FOODCHEM.2014.10.098.

Alarcon-Barrera, Karina S., Daniela S. Armijos-Montesinosa, Marilyn García-Tenesacaa, Gabriel Iturralde, Tatiana Jaramilo-Vivancoc, Maria G. Granda-Albujad, Francesca Giampierie, and Jose M. Alvarez-Suarez. 2018. "Wild Andean Blackberry (*Rubus glaucus* Benth) and Andean Blueberry (*Vaccinium floribundum* Kunth) from the Highlands of Ecuador: Nutritional Composition and Protective Effect on Human Dermal Fibroblasts against Cytotoxic Oxidative Damage." *Journal of Berry Research* 8 (3): 223–36. doi: 10.3233/JBR-180316.

Azofeifa, Gabriela, Silvia Quesada, Frederic Boudard, Marion Morena, Jean Paul Cristol, Ana M. Pérez, Fabrice Vaillant, and Alain Michel. 2013. "Antioxidant and Anti-Inflammatory *in Vitro* Activities of Phenolic Compounds from Tropical Highland Blackberry (*Rubus adenotrichos*)." *Journal of Agricultural and Food Chemistry* 61 (24): 5798–5804. doi: 10.1021/jf400781m.

Azofeifa, Gabriela, Silvia Quesada, Laura Navarro, Olman Hidalgo, Karine Portet, Ana M. Pérez, Fabrice Vaillant, Patrick Poucheret, and Alain Michel. 2016. "Hypoglycaemic, Hypolipidaemic and Antioxidant Effects of Blackberry Beverage Consumption in Streptozotocin-Induced Diabetic Rats." *Journal of Functional Foods* 26: 330–37. doi: 10.1016/j.jff.2016.08.007.

Azofeifa, Gabriela, Silvia Quesada, Ana M. Pérez, Fabrice Vaillant, and Alain Michel. 2015. "Pasteurization of Blackberry Juice Preserves Polyphenol-Dependent Inhibition for Lipid Peroxidation and Intracellular Radicals." *Journal of Food Composition and Analysis* 42: 56–62. doi: 10.1016/J.JFCA.2015.01.015.

Baby, Bincy, Priya Antony, and Ranjit Vijayan. 2018. "Antioxidant and Anticancer Properties of Berries." *Critical Reviews in Food Science and Nutrition* 58 (15): 2491–2507. doi: 10.1080/10408398.2017.1329198.

Beristain-Bauza, Silvia Del Carmen, Paola Hernández-Carranza, Teresa Soledad Cid-Pérez, Raúl Ávila-Sosa, Irving Israel Ruiz-López, and Carlos Enrique Ochoa-Velasco. 2019. "Antimicrobial Activity of Ginger (*Zingiber officinale*) and Its Application in Food Products."

Food Reviews International 35 (5): 407–26. doi: 10.1080/87559129.2019.1573829.

Betta, Fabiana Della, Priscila Nehring, Siluana Katia Tischer Seraglio, Mayara Schulz, Andressa Camargo Valese, Heitor Daguer, Luciano Valdemiro Gonzaga, Roseane Fett, and Ana Carolina Oliveira Costa. 2018. "Phenolic Compounds Determined by LC-MS/MS and *In Vitro* Antioxidant Capacity of Brazilian Fruits in Two Edible Ripening Stages." *Plant Foods for Human Nutrition* 73 (4). Plant Foods for Human Nutrition: 302–7. doi: 10.1007/s11130-018-0690-1.

Borochov-Neori, Hamutal, Sylvie Judeinstein, Amnon Greenberg, Nina Volkova, Mira Rosenblat, and Michael Aviram. 2015. "Antioxidant and Antiatherogenic Properties of Phenolic Acid and Flavonol Fractions of Fruits of 'Amari' and 'Hallawi' Date (*Phoenix Dactylifera* L.) Varieties." *Journal of Agricultural and Food Chemistry* 63 (12): 3189–95. doi: 10.1021/jf506094r.

Cassidy, Aedín. 2018. "Berry Anthocyanin Intake and Cardiovascular Health." *Molecular Aspects of Medicine* 61: 76–82. doi: 10.1016/j.mam.2017.05.002.

Chaves-Silva, Samuel, Adolfo Luís Dos Santos, Antonio Chalfun-Júnior, Jian Zhao, Lázaro E.P. Peres, and Vagner Augusto Benedito. 2018. "Understanding the Genetic Regulation of Anthocyanin Biosynthesis in Plants – Tools for Breeding Purple Varieties of Fruits and Vegetables." *Phytochemistry* 153: 11–27. doi: 10.1016/j.phytochem.2018.05.013.

Cheng, Pengfei, Jingxia Wang, and Weihua Shao. 2016. "Monounsaturated Fatty Acid Intake and Stroke Risk: A Meta-Analysis of Prospective Cohort Studies." *Journal of Stroke and Cerebrovascular Diseases* 25 (6): 1326–34. doi: 10.1016/J.JSTROKECEREBROVASDIS.2016.02.017.

Cömert, Ezgi Doğan, and Vural Gökmen. 2018. "Evolution of Food Antioxidants as a Core Topic of Food Science for a Century." *Food Research International* 105: 76–93. doi: 10.1016/j.foodres.2017.10.056.

Contessa, Cecilia, Maria Gabriella Mellano, Gabriele Loris Beccaro, Annalisa Giusiano, and Roberto Botta. 2013. "Total Antioxidant Capacity and Total Phenolic and Anthocyanin Contents in Fruit Species

Grown in Northwest Italy." *Scientia Horticulturae* 160: 351–57. doi: 10.1016/j.scienta.2013.06.019.

Costa, Francelina, Maria Lurdes de Baeta, David Saraiva, Manuel Teixeira Verissimo, and Fernando Ramos. 2011. "Evolution of Mineral Contents in Tomato Fruits during the Ripening Process after Harvest." *Food Analytical Methods* 4 (3): 410–15. doi: 10.1007/s12161-010-9179-8.

Dabbou, Samia, Samira Maatallah, Antonella Castagna, Monia Guizani, Wala Sghaeir, Hichem Hajlaoui, and Annamaria Ranieri. 2017. "Carotenoids, Phenolic Profile, Mineral Content and Antioxidant Properties in Flesh and Peel of *Prunus persica* Fruits during Two Maturation Stages." *Plant Foods for Human Nutrition* 72 (1). *Plant Foods for Human Nutrition*: 103–10. doi: 10.1007/s11130-016-0585-y.

Das, Quail, Md Rashedul Islam, Massimo F. Marcone, Keith Warriner, and Moussa S. Diarra. 2017. "Potential of Berry Extracts to Control Foodborne Pathogens." *Food Control* 73: 650–62. doi: 10.1016/j.foodcont.2016.09.019.

Fang, Jim. 2015. "Classification of Fruits Based on Anthocyanin Types and Relevance to Their Health Effects." *Nutrition* 31 (11–12): 1301–6. doi: 10.1016/j.nut.2015.04.015.

Figueroa-Méndez, Rodrigo, and Selva Rivas-Arancibia. 2015. "Vitamin C in Health and Disease: Its Role in the Metabolism of Cells and Redox State in the Brain." *Frontiers in Physiology* 6: 397. doi: 10.3389/fphys.2015.00397.

Gancel, Anne Laure, Aurélien Feneuil, Oscar Acosta, Ana Mercedes Pérez, and Fabrice Vaillant. 2011. "Impact of Industrial Processing and Storage on Major Polyphenols and the Antioxidant Capacity of Tropical Highland Blackberry (*Rubus adenotrichus*)." *Food Research International* 44 (7): 2243–51. doi: 10.1016/j.foodres.2010.06.013.

Garzón, G. a., K. M. Riedl, and S. J. Schwartz. 2009. "Determination of Anthocyanins, Total Phenolic Content, and Antioxidant Activity in Andes Berry (*Rubus glaucus* Benth)." *Journal of Food Science* 74 (3): 227–32. doi: 10.1111/j.1750-3841.2009.01092.x.

Goh, Joanna Xuan Hui, Loh Teng-Hern Tan, Joo Kheng Goh, Kok Gan Chan, Priyia Pusparajah, Learn-Han Lee, and Bey-Hing Goh. 2019.

"Nobiletin and Derivatives: Functional Compounds from Citrus Fruit Peel for Colon Cancer Chemoprevention." *Cancers* 11 (6): 867. doi: 10.3390/cancers11060867.

Hajaji, Soumaya, Mohamed Amine Jabri, Ines Sifaoui, Atteneri López-Arencibia, María Reyes-Batlle, Fatma B'chir, Basilio Valladares, José E. Pinero, Jacob Lorenzo-Morales, and Hafidh Akkari. 2017. "Amoebicidal, Antimicrobial and *in Vitro* ROS Scavenging Activities of Tunisian *Rubus ulmifolius* Schott, Methanolic Extract." *Experimental Parasitology* 183: 224–30. doi: 10.1016/j.exppara.2017.09.013.

Hassimotto, Neuza Mariko Aymoto, Marcia Da Silva Pinto, and Franco Maria Lajolo. 2008. "Antioxidant Status in Humans after Consumption of Blackberry (*Rubus fruticosus* L.) Juices with and without Defatted Milk." *Journal of Agricultural and Food Chemistry* 56 (24): 11727–33. doi: 10.1021/jf8026149.

Hidalgo, Gádor-Indra, and María Pilar Almajano. 2017. "Red Fruits: Extraction of Antioxidants, Phenolic Content, and Radical Scavenging Determination: A Review." *Antioxidants* 6: 1–27. doi: 10.3390/antiox 6010007.

Horvitz, Sandra, Diana Chanaguano, and Iñigo Arozarena. 2017. "Andean Blackberries (*Rubus glaucus* Benth) Quality as Affected by Harvest Maturity and Storage Conditions." *Scientia Horticulturae* 226: 293–301. doi: 10.1016/j.scienta.2017.09.002.

Hui, Y. H., J´ozsef Barta, M. Pilar Cano, Todd W. Gusek, Jiwan S. Sidhu, and Nirmal K. Sinha. 2010. *Handbook of Fruits and Fruit Processing*. Iowa: Blackwell Publishing.

Jaakola, Laura. 2013. "New Insights into the Regulation of Anthocyanin Biosynthesis in Fruits." *Trends in Plant Science* 18 (9): 477–83. doi: 10.1016/j.tplants.2013.06.003.

Jakobek, Lidija, Marijan Šeruga, Bernarda Šeruga, Ivana Novak, and Martina Medvidović-Kosanović. 2009. "Phenolic Compound Composition and Antioxidant Activity of Fruits of *Rubus* and *Prunus* Species from Croatia." *International Journal of Food Science and Technology* 44 (4): 860–68. doi: 10.1111/j.1365-2621.2009.01920.x.

Joseph, Shama V, Indika Edirisinghe, and Britt M. Burton-Freeman. 2016. "Fruit Polyphenols: A Review of Anti-Inflammatory Effects in Humans." *Critical Reviews in Food Science and Nutrition* 56 (3): 419–44. doi: 10.1080/10408398.2013.767221.

Kamiloglu, Senem, Ayca Ayfer Pasli, Beraat Ozcelik, John Van Camp, and Esra Capanoglu. 2015. "Influence of Different Processing and Storage Conditions on *in Vitro* Bioaccessibility of Polyphenols in Black Carrot Jams and Marmalades." *Food Chemistry* 186. doi: 10.1016/j.foodchem. 2014.12.046.

Koczka, Noémi, Éva Stefanovits-Bányai, and Eniko Prokaj. 2018. "Element Composition, Total Phenolics and Antioxidant Activity of Wild and Cultivated Blackberry (*Rubus fruticosus* L.) Fruits and Leaves during the Harvest Time." *Notulae Botanicae Horti Agrobotanici Cluj-Napoca* 46 (2): 563–69. doi: 10.15835/nbha46210993.

Kostecka-Gugała, Anna, Iwona Ledwożyw-Smoleń, Joanna Augustynowicz, Gabriela Wyżgolik, Michał Kruczek, and Paweł Kaszycki. 2015. "Antioxidant Properties of Fruits of Raspberry and Blackberry Grown in Central Europe." *Open Chemistry* 13 (1): 1313–25. doi: 10.1515/chem-2015-0143.

Landete, J. M. 2011. "Ellagitannins, Ellagic Acid and Their Derived Metabolites: A Review about Source, Metabolism, Functions and Health." *Food Research International* 44 (5). Elsevier Ltd: 1150–60. doi: 10.1016/j.foodres.2011.04.027.

Larrosa, Mar, María T. García-Conesa, Juan C. Espín, and Francisco A. Tomás-Barberán. 2010. "Ellagitannins, Ellagic Acid and Vascular Health." *Molecular Aspects of Medicine* 31 (6). Elsevier Ltd: 513–39. doi: 10.1016/j.mam.2010.09.005.

Lee, Jungmin, Michael Dossett, and Chad E. Finn. 2012. "*Rubus* Fruit Phenolic Research: The Good, the Bad, and the Confusing." *Food Chemistry* 130 (4). Elsevier Ltd: 785–96. doi: 10.1016/j.foodchem. 2011.08.022.

Leterm, Pascal, Andre Buldgen, Fernando Estrada, and Angela M. Londoño. 2006. "Mineral Content of Tropical Fruits and Unconventional Foods of the Andes and the Rain Forest of Colombia." *Food Chemistry* 95: 644–652.

Ljevar, Ana, Natka Ćurko, Marina Tomašević, Kristina Radošević, Višnja Gaurina Srček, and Karin Kovačević Ganić. 2016. "Phenolic Composition, Antioxidant Capacity and *in Vitro* Cytotoxicity Assessment of Fruit Wines." *Food Technology and Biotechnology* 54 (2): 145–155. doi: 10.17113/fb.54.02.16.4208.

Ma, Xiaowei, Hongxia Wu, Liqin Liu, Quansheng Yao, Songbiao Wang, Rulin Zhan, Shanshan Xing, and Yigang Zhou. 2011. "Polyphenolic Compounds and Antioxidant Properties in Mango Fruits." *Scientia Horticulturae* 129 (1): 102–7. doi: 10.1016/j.scienta.2011.03.015.

Maleki, Soheila J., Jesus F. Crespo, and Beatriz Cabanillas. 2019. "Anti-Inflammatory Effects of Flavonoids." *Food Chemistry* 299: 125124. doi: 10.1016/j.foodchem.2019.125124.

Mertz, Christian, Veronique Cheynier, Ziya Günata, and Pierre Brat. 2007. "Analysis of Phenolic Compounds in Two Blackberry Species (*Rubus glaucus* and *Rubus adenotrichus*) by High-Performance Liquid Chromatography with Diode Array Detection and Electrospray Ion Trap Mass Spectrometry." *Journal of Agricultural and Food Chemistry* 55 (21): 8616–24. doi: 10.1021/jf071475d.

Milivojević, J., V. Maksimović, M. Nikolić, J. Bogdanović, R. Maletić, and D. Milatović. 2011. "Chemical and Antioxidant Properties of Cultivated and Wild Fragaria and Rubus Berries." *Journal of Food Quality* 34 (1): 1–9. doi: 10.1111/j.1745-4557.2010.00360.x.

Monforte, Maria Teresa, Antonella Smeriglio, Maria Paola Germanò, Simona Pergolizzi, Clara Circosta, and Enza Maria Galati. 2018. "Evaluation of Antioxidant, Antiinflammatory, and Gastroprotective Properties of *Rubus fruticosus* L. Fruit Juice." *Phytotherapy Research* 32 (7): 1404–14. doi: 10.1002/ptr.6078.

Moraes, Caroline da Silva, Francisco Odencio Rodrigues de Oliveira Junior, Gustavo Masson, Karina Mastropasqua Rebello, Lívia de Oliveira Santos, Narayana Fazolini P. Bastos, and Rozana Côrte-Real Faria. 2013. *Métodos Experimentais no Estudo de Proteínas*. Rio de Janeiro: IOC.

Morales, Patricia, Isabel C. F. R. Ferreira, Ana Maria Carvalho, Virginia Fernández-Ruiz, Ma S. O. S. C. C. Sánchez-Mata, Montaña Cámara, Ramón Morales, and Javier Tardío. 2013a. "Wild Edible Fruits as a Potential Source of Phytochemicals with Capacity to Inhibit Lipid Peroxidation." *European Journal of Lipid Science and Technology* 115 (2): 176–85. doi: 10.1002/ejlt.201200162.

Mulvihill, Erin E., and Murray W. Huff. 2010. "Antiatherogenic Properties of Flavonoids: Implications for Cardiovascular Health." *Canadian Journal of Cardiology* 26: 17–21. doi: 10.1016/s0828-282x(10)71056-4.

Nowak, Adriana, Michał Sójka, Elżbieta Klewicka, Lidia Lipińska, Robert Klewicki, and Krzysztof Kołodziejczyk. 2017. "Ellagitannins from *Rubus idaeus* L. Exert Geno- and Cytotoxic Effects against Human Colon Adenocarcinoma Cell Line Caco-2." *Journal of Agricultural and Food Chemistry* 65: 2947–2955. doi:10.1021/acs.jafc.6b05387.

Ogawa, Kenjirou, Hiroyuki Sakakibara, Rei Iwata, Takeshi Ishii, Tsutomu Sato, Toshinao Goda, Kayoko Shimoi, and Shigenori Kumazawa. 2008. "Anthocyanin Composition and Antioxidant Activity of the Crowberry (*Empetrum nigrum*) and Other Berries." *Journal of Agricultural and Food Chemistry* 56 (12): 4457–62. doi: 10.1021/jf800406v.

Ooi, Esther M. M., Gerald F. Watts, Theodore W. K. Ng, and P. H. R. Barrett. 2015. "Effect of Dietary Fatty Acids on Human Lipoprotein Metabolism: A Comprehensive Update." *Nutrients* 7: 4416–25. doi: 10.3390/nu7064416.

Peraza-Padilla, W., and M. Orozco-Aceves. 2018. "Plant-Parasitic Nematodes Associated with Blackberry (*Rubus adenotrichus* SCHLTDL.) Plantations in Costa Rica." *Nematropica* 48 (2): 145–54. https://pdfs.semanticscholar.org/5cb7/717bbc6c2eb20b0f7680401148e 5468c453d.pdf?_ga=2.192244842.1306894989.1566927640-158548 2673.1566927640.

Pereira-Netto, Adaucto B. 2018. "Tropical Fruits as Natural, Exceptionally Rich, Sources of Bioactive Compounds." *International Journal of Fruit Science* 18 (3): 231–42. doi: 10.1080/15538362.2018.1444532.

Pinto, Teresa, Alice Vilela, Andreia Pinto, Fernando M. Nunes, Fernanda Cosme, and Rosário Anjos. 2018. “Influence of Cultivar and of Conventional and Organic Agricultural Practices on Phenolic and Sensory Profile of Blackberries (Rubus Fruticosus).” *Journal of the Science of Food and Agriculture* 98 (12): 4616–24. doi: 10.1002/jsfa.8990.

Plessi, M., D. Bertelli, and a. Albasini. 2007. “Distribution of Metals and Phenolic Compounds as a Criterion to Evaluate Variety of Berries and Related Jams.” *Food Chemistry* 100 (1): 419–27. doi: 10.1016/j.foodchem.2005.09.018.

Radovanović, Blaga C., Ana S Milenković Andelković, Aleksandra B. Radovanović, and Marko Z. Andelković. 2013. “Antioxidant and Antimicrobial Activity of Polyphenol Extracts from Wild Berry Fruits Grown in Southeast Serbia.” *Tropical Journal of Pharmaceutical Research* 12 (5): 813–19. doi: 10.4314/tjpr.v12i5.23.

Rahman, Mohammed Shamim, Andrew J. Murphy, and Kevin J. Woollard. 2017. “Effects of Dyslipidaemia on Monocyte Production and Function in Cardiovascular Disease.” *Nature Reviews Cardiology* 14 (7): 387–400. doi: 10.1038/nrcardio.2017.34.

Ramful, Deena, Evelyne Tarnus, Okezie I. Aruoma, Emmanuel Bourdon, and Theeshan Bahorun. 2011. “Polyphenol Composition, Vitamin C Content and Antioxidant Capacity of *Mauritian citru*s Fruit Pulps.” *Food Research International* 44 (7): 2088–99. doi: 10.1016/J.FOODRES.2011.03.056.

Redondo, German L. Madrigal, Rolando Vargas Zúñiga, Gustavo Carazo Berrocal, Nils Arguedas Ramirez, Lidiette Fonseca González, and Jorge Campos. 2017. “*Rubus adenotrichus* Fruit Extracts Phytochemical Characterization and Antioxidant Power Evaluation for Dermocosmetic Formulations.” *International Journal of Phytocosmetics and Natural Ingredients* 4 (5): 1–10. doi: 10.15171/ijpni.2017.05.

Reidel, Rose Vanessa Bandeira, Bernardo Melai, Pierluigi Cioni, Guido Flamini, and Luisa Pistelli. 2016. “Aroma Profile of *Rubus ulmifolius* Flowers and Fruits during Different Ontogenetic Phases.” *Chemistry and Biodiversity* 13 (12): 1776–84. doi: 10.1002/cbdv.201600170.

Riaz, Muhammad, Mansoor Ahmad, and Najmur Rahman. 2011. "Antimicrobial Screening of Fruit, Leaves, Root and Stem of *Rubus fruticosus*." *Journal of Medicinal Plants Research* 5 (24): 5920–24. http://www.academicjournals.org/JMPR/abstracts/abstracts/abstracts2011/30Oct/Riaz et al.htm.

Rufino, Maria Do Socorro M., Ricardo E. Alves, Edy S. de Brito, Jara Pérez-Jiménez, Fulgencio Saura-Calixto, and Jorge Mancini-Filho. 2010. "Bioactive Compounds and Antioxidant Capacities of 18 Non-Traditional Tropical Fruits from Brazil." *Food Chemistry* 121 (4): 996–1002. doi: 10.1016/j.foodchem.2010.01.037.

Ruiz-Rodríguez, Brígida María, Concepción Sánchez-Moreno, Begoña De Ancos, María De Cortes Sánchez-Mata, Virginia Fernández-Ruiz, Montaña Cámara, and Javier Tardío. 2014. "Wild *Arbutus unedo* L. and *Rubus ulmifolius* Schott Fruits Are Underutilized Sources of Valuable Bioactive Compounds with Antioxidant Capacity." *Fruits* 69 (6): 435–48. doi: 10.1051/fruits/2014035.

Sangiovanni, Enrico, Urska Vrhovsek, Giuseppe Rossoni, Elisa Colombo, Cecilia Brunelli, Laura Brembati, Silvio Trivulzio, et al. 2013. "Ellagitannins from Rubus Berries for the Control of Gastric Inflammation: *In Vitro* and *In Vivo* Studies." *PLoS ONE* 8 (8): 1–12. doi: 10.1371/journal.pone.0071762.

Sariburun, Esra, Saliha Şahin, Cevdet Demir, Cihat Türkben, and Vildan Uylaşer. 2010. "Phenolic Content and Antioxidant Activity of Raspberry and Blackberry Cultivars." *Journal of Food Science* 75 (4): 328–35. doi: 10.1111/j.1750-3841.2010.01571.x.

Schulz, Mayara, and Josiane Freitas Chim. 2019. "Nutritional and Bioactive Value of *Rubus* Berries." *Food Bioscience* 31 (100438): 1–14. doi: 10.1016/j.fbio.2019.100438.

Schulz, Mayara, Siluana Katia Tischer Seraglio, Fabiana Della Betta, Priscila Nehring, Andressa Camargo Valese, Heitor Daguer, Luciano Valdemiro Gonzaga, Ana Carolina Oliveira Costa, and Roseane Fett. 2019. "Blackberry (*Rubus ulmifolius* Schott): Chemical Composition, Phenolic Compounds and Antioxidant Capacity in Two Edible Stages."

Food Research International 122: 627–34. doi: 10.1016/j.foodres.2019.01.034.

Seraglio, Siluana Katia Tischer, Mayara Schulz, Priscila Nehring, Fabiana Della Betta, Andressa Camargo Valese, Heitor Daguer, Luciano Valdemiro Gonzaga, Roseane Fett, and Ana Carolina Oliveira Costa. 2018. "Nutritional and Bioactive Potential of Myrtaceae Fruits during Ripening." *Food Chemistry* 239: 649–56. doi: 10.1016/j.foodchem.2017.06.118.

Shahidi, Fereidoon, and Priyatharini Ambigaipalan. 2015. "Phenolics and Polyphenolics in Foods, Beverages and Spices: Antioxidant Activity and Health Effects - A Review." *Journal of Functional Foods* 18. Elsevier Ltd: 820–97. doi: 10.1016/j.jff.2015.06.018.

Silva, Liliana Primo Da, Eliana Pereira, Tânia C.S.P. Pires, Maria José Alves, Olívia R. Pereira, Lillian Barros, and Isabel C.F.R. Ferreira. 2019. "*Rubus ulmifolius* Schott Fruits: A Detailed Study of Its Nutritional, Chemical and Bioactive Properties." *Food Research International* 119: 34–43. https://doi.org/10.1016/j.foodres.2019.01.052.

Soto, Marvin, Ana M. Perez, María del Milagro Cerdas, Fabrice Vaillant, and Óscar Acosta. 2019. "Physicochemical Characteristics and Polyphenolic Compounds of Cultivated Blackberries in Costa Rica." *Journal of Berry Research* 9 (2): 283–96. doi: 10.3233/JBR-180353.

Souza, Vanessa Rios de, Patrícia Aparecida Pimenta Pereira, Thais Lomônaco Teodoro da Silva, Luiz Carlos de Oliveira Lima, Rafael Pio, and Fabiana Queiroz. 2014. "Determination of the Bioactive Compounds, Antioxidant Activity and Chemical Composition of Brazilian Blackberry, Red Raspberry, Strawberry, Blueberry and Sweet Cherry Fruits." *Food Chemistry* 156: 362–68. doi: 10.1016/j.foodchem.2014.01.125.

Taiz, L., and E. Zeiger. 2009. *Fisiologia Vegetal*. 4th ed. Porto Alegre: Artmed.

Türkben, Cihat, Esra Sariburun, Cevdet Demir, and Vildan Uylaşer. 2010. "Effect of Freezing and Frozen Storage on Phenolic Compounds of

Raspberry and Blackberry Cultivars." *Food Analytical Methods* 3 (3): 144–53. doi: 10.1007/s12161-009-9102-3.

Vasco, Catalina, Kaisu Riihinen, Jenny Ruales, and Afaf Kamal-Eldin. 2009. "Phenolic Compounds in Rosaceae Fruits from Ecuador." *Journal of Agricultural and Food Chemistry* 57 (4): 1204–12. doi: 10.1021/jf802656r.

Velde, Franco Van de, Debora Esposito, Mary H. Grace, María E. Pirovani, and Mary A. Lila. 2019. "Anti-Inflammatory and Wound Healing Properties of Polyphenolic Extracts from Strawberry and Blackberry Fruits." *Food Research International* 121: 453–62. doi: 10.1016/j.foodres.2018.11.059.

Velde, Franco Van de, Mary H. Grace, Debora Esposito, María Élida Pirovani, and Mary Ann Lila. 2016. "Quantitative Comparison of Phytochemical Profile, Antioxidant, and Anti-Inflammatory Properties of Blackberry Fruits Adapted to Argentina." *Journal of Food Composition and Analysis* 47: 82–91. doi: 10.1016/j.jfca.2016.01.008.

Vergara, Manuel Fernando, Jessica Vargas, and John Fabio Acuña. 2016. "Physicochemical Characteristics of Blackberry (*Rubus glaucus* Benth.) Fruits from Four Production Zones of Cundinamarca, Colombia." *Agronomía Colombian* 34 (3): 336–45. http://dx.doi.org/10.15446/agron.colomb.v34n3.62755.

Yang, Ji Won, and Il Sook Choi. 2017. "Comparison of the Phenolic Composition and Antioxidant Activity of Korean Black Raspberry, Bokbunja, (*Rubus coreanus* Miquel) with Those of Six Other Berries." *CyTA - Journal of Food* 15 (1): 110–117. doi: 10.1080/19476337.2016.1219390.

Yeung, Andy Wai Kan, Nikolay T. Tzvetkov, Gokhan Zengin, Dongdong Wang, Suowen Xu, Goranka Mitrovic, Mladen Brncic, et al. 2019. "The Berries on the Top." *Journal of Berry Research* 9 (1): 125–39. doi: 10.3233/JBR-180357.

Yilmaz, Kadir Ugurtan, Yasar Zengin, Sezai Ercisli, Sedat Serce, Kazim Gunduz, Memnune Sengul, and Bayram Murat Asma. 2009. "Some Selected Physico-Chemical Characteristics of Wild and Cultivated Blackberry Fruits (*Rubus fruticosus* L.) from Turkey." *Romanian*

Biotechnological Letters 14 (1): 4152–63. http://www.rombio.eu/rbl1vol14/1-5/lucr-10-Kadir-Yilmaz-bt.pdf.

Zambrano, Alexandra, Rosa Raybaudi-Massilia, Francisco Arvelo, and Felipe Sojo. 2018. "Cytotoxic and Antioxidant Properties *in Vitro* of Functional Beverages Based on Blackberry (*Rubus glaucus* B.) and Soursop (*Annona muricata* L.) Pulps." *Functional Foods in Health and Disease* 8 (11): 531–47. doi: 10.31989/ffhd.v8i11.541.

Zhang, Huijuan, Hongna Wang, Xinran Cao, and Jing Wang. 2018. "Preparation and Modification of High Dietary Fiber Flour: A Review." *Food Research International* 113: 24–35. doi: 10.1016/j.foodres.2018.06.068.

Zhang, Lijuan, Carlos Virgous, and Hongwei Si. 2019. "Synergistic Anti-Inflammatory Effects and Mechanisms of Combined Phytochemicals." *Journal of Nutritional Biochemistry* 69: 19–30. doi: 10.1016/j.jnutbio.2019.03.009.

Zia-Ul-Haq, Muhammad, Muhammad Riaz, Vincenzo De Feo, Hawa Z E Jaafar, and Marius Moga. 2014. "*Rubus fruticosus* L.: Constituents, Biological Activities and Health Related Uses." *Molecules* 19 (8): 10998–29. doi: 10.3390/molecules190810998.

In: Rubus: An Overview
Editor: Davet Chabot

ISBN: 978-1-53617-376-5

Chapter 2

RUBUS: BIOACTIVE COMPOUNDS AND THEIR NEUROPROTECTIVE EFFECT

Brenda Pérez-Grijalva*[*]*, PhD,
Diana E. Leyva-Daniel, PhD
and Rosalva Mora-Escobedo, PhD
Departamento de Ingeniería Bioquímica, Escuela Nacional de Ciencias Biológicas, Instituto Politécnico Nacional,
Ciudad de México, CDMX, México

ABSTRACT

The bioactive compounds present in the genus *Rubus,* mainly, phenolic compounds such as flavonoids (anthocyanins, flavanols and phenolic acids) and ellagitannins, have garnered special attention due to the health benefits of their consumption. However, their biological activity is related with their absorption and metabolism. The purpose of this chapter is to summarize the available scientific information of the bioactive compounds in *Rubus* (blackberries and raspberries) and their bioavalibility, as well as their effect on intestinal microbiota and tissue distribution. Similarly, we review the role of the phenolic compounds in *Rubus* berries

[*] Corresponding Author's Email: braq82@hotmail.com.

in the prevention and treatment of neurodegenerative disorders. Experimental studies suggest that anthocyanins of these fruits improve cognition and human health.

Keywords: *Rubus*, bioactive compounds, anthocyanins, nutraceutical behavior, microbiota, neuroprotective effect

INTRODUCTION

The *Rubus* genus groups plant species belong to the Rosaceae family, subfamily Rosoideae. Members of this genus are native to six continents at elevations from sea level to 4500 m and over 740 species have been described worldwide (Hummer, 2010; Schulz and Chim, 2019; Ahmad et al., 2015). Species of the genus *Rubus* are characterized by having stems like roses and are often called brambles. The growth habit varies from erect to trailing types; some plants are climbing shrubs (*R. cissoides Cunn.*), others as *R. chamaemorus* L. take dwarf alpine forms that spread by underground stolons. Most species are deciduous, but some are evergreen (Hummer, 1996).

Ripe fruit color includes white, yellow, orange, red, purple, and black. Some fruits have orange drupelets with purple-black at the stylar end. Raspberry-type fruits consist of an aggregate of drupelets that pull away from the floral receptacle while blackberry drupelets adhere to it. In blackberries, the drupelets and receptacle separate from the pedicel upon harvest (Hummer, 1996). Berries, especially members of the Rosaceae (strawberry, raspberry, blackberry) and Ericaceae (blueberry, cranberry) families, are the best dietary sources of bioactive compounds. They have delicious taste and flavor, besides having economic importance. Given the antioxidant properties of the bioactive compounds they contain, berries represent a great opportunity for nutritionists and food technologists to use bioactive compounds as ingredients in functional foods (Skrovankova, et al., 2015).

Plants synthesize a variety of medically active phytochemicals; most are derivates of alkaloids, phenolics, terpenoids, and glycosides. Phenolic derivates are the most active biochemical motif in *Rubus* plants for ethnomedicinal applications. *Rubus* phenolics include the flavonoids, and potent in vitro antioxidants, including compounds such as flavones, isoflavones, flavanones, catechins, and the red, blue, and purple pigments known as anthocyanins (Määttä-Riihinen et al., 2004). These compounds, either individual or combined, are responsible for various health benefits of *Rubus* genus, such as the prevention of inflammation disorders, cardiovascular diseases, and protective effects to lower the risk of various cancers (Skrovankova, et al., 2015). In this chapter, bioactive compounds of commonly consumed fruits of *Rubus* genus are described as well as the factors influencing their antioxidant capacity and health benefits, emphasizing the neuroprotective effect that these compounds could have.

1. Bioactive Compounds

Berries are simple fruits that mostly belong to the Rosaceae and Ericaceae families. Rosaceae family is commonly known as brambles. Rosaceae and Ericaceae families include strawberry (*Fragaria ananassa*), blackberry (*Rubus fruticosus*), raspberry (*Rubus idaeus*), and cranberry (Vaccinium macrocarpon), among others (See Table 1) (Olas, 2018). The genus *Rubus*, includes blackberry and raspberry; these berries are widely distributed worldwide (Schulz and Chim, 2019; Ahmad et al., 2015).

For the past years, consumption of fresh and processed (i.e., drinks, ice-cream, jams) berries has increased as a result of several studies that reveal they constitute a relevant source of bioactive compounds (Seeram et al., 2006; Manganaris et al., 2014). Among these substances are phenolic compounds, vitamins, and carotenoids.

Within the micronutrients of relevance that have been reported in berries and brambles, there are minerals such as calcium, phosphorus, magnesium, manganese, potassium, copper, and sodium (Olas, 2018; Hariram et al., 2014; Schulz and Chim, 2019). In addition to phenolic compounds, berries also contain vitamins A, E, and C, the latter being predominant in berries as blackcurrants (120–215 mg vitamin C/100 g) and sea buckthorn (600 mg vitamin C/100 g) (Olas, 2016).

Table 1. Botanical classification of berries

Family	Examples
Rosaceae	Black chokeberry (*Aronia melanocarpa*)
	Strawberry (*Fragaria ananassa*)
	Red raspberry (*Rubus idaeus*)
	Black raspberry (*Rubus occidentalis*)
	Blackberry (*Rubus fruticosus*)
	Loganberry (*Rubus loganobaccus*)
	Cloudberry (*Rubus chamaemorus*)
Ericaceae	Cranberry (*Vaccinium macrocarpon*)
	Rabbit eye blueberries (*Vaccinium ashei*)
	Bilberry (*Vaccinium myrtillus*)
	Lowbush blueberry (*Vaccinium angustifolium*)
	Highbush blueberry (*Vaccinium corymbosum*)
Others	Blackcurrants (*Ribes nigrum*; family: *Grossulariaceae*)
	Gooseberry (*Ribes uva-crispa*; family: *Grossulariaceae*)
	Sea buckthorn (*Elaeagnus rhamnoides* (L.); family: *Elaeagnaceae*)
	Grapes (*Vitis*; family: *Vitaceae*)

1.1. Phenolic Compounds

Phenolic compounds are secondary metabolites with an aromatic ring consisting of one or more substitutes of hydroxyl (-OH), methoxyl (-OCH_3), and glucoside groups (Figure 1) (Manganaris et al., 2014; Crozier et al., 2006). These compounds contribute to the color (i.e., anthocyanins) and flavor (i.e., tannins) of berries (Paredes-López et al., 2019). Additionally, they play key roles in plants, such as pollinator attraction, pigmentation, antioxidant activity, and protection against UV radiation.

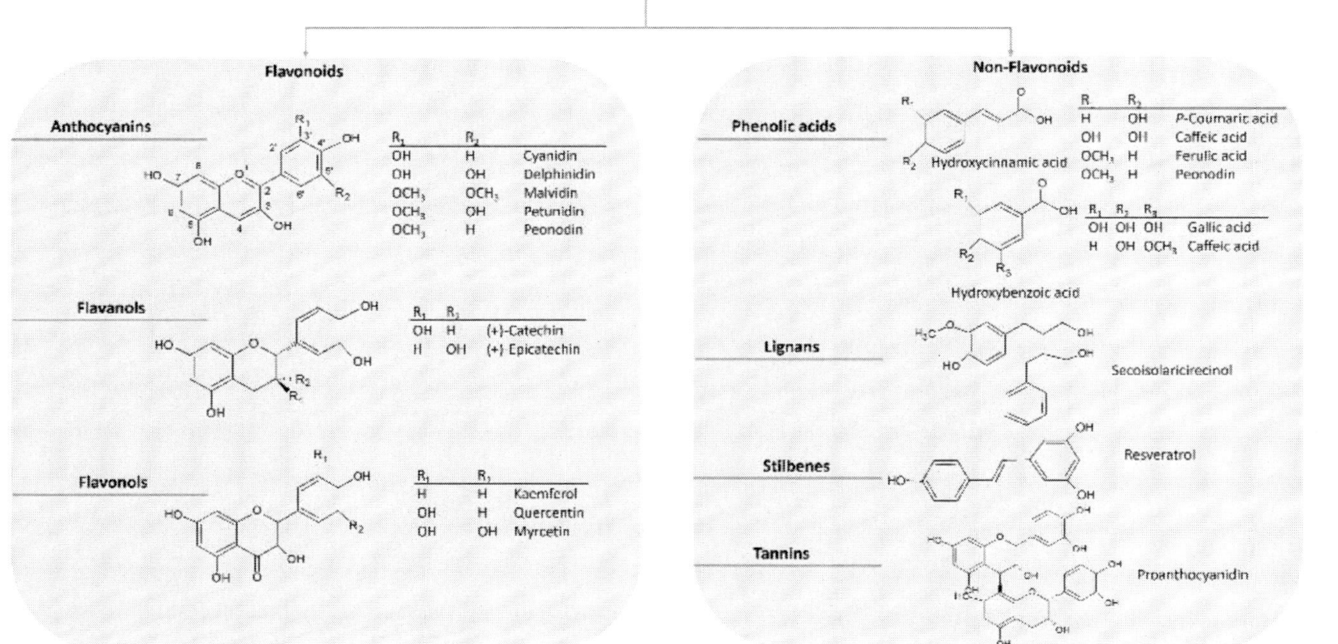

Figure 1. Classification of Phenolic compounds.

The concentration of these compounds in fruits is likely due to several factors, such as species, maturity stage, soil, and weather (Schulz and Chim, 2019). Phenolic compounds are attributed different biological activities, including antioxidant, anti-inflammatory, antibacterial, and antithrombotic, as well as a neuroprotective effect (Tungmunnithum et al., 2018; Elisha et al., 2016; Kelsey et al., 2011; Gwiazdon and Biesaga, 2017). The most relevant phenolic compounds reported in berries are phenolic acids, flavonoids (flavanols, flavonols, anthocyanins), tannins, and lignans (Figure 2) (Manganaris et al., 2014; Paredes-López et al., 2010). Table 2 shows the total phenol content in some berries. The highest phenolic concentrations are found in raspberry, (2494 mg GAE/100 g fresh weight, FW) and blackberry (2349 mg GAE/100 g FW).

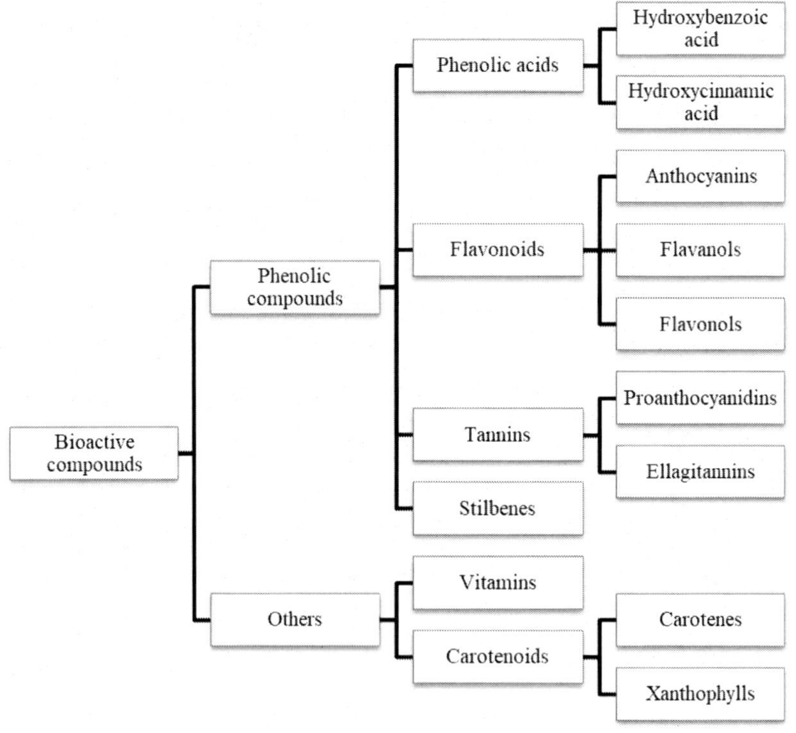

Figure 2. Bioactive compounds in berries.

Table 2. Total phenolic and monomeric anthocyanin content in some berries

Berry	Total phenolic content (mg GAE/100 g FW)	Total monomeric anthocyanins (mg C3GE/100 g FW)	References
Blackberry (*Rubus fruticosus*)	545–2349	89–211	(Szajdek and Borowska, 2008; Peña-Sanhueza et al., 2017; Paredes-López et al., 2010; Schulz and Chim, 2019)
Red raspberry (*Rubus idaeus*)	142–553	14.7–49.4	
Blueberry (*Vaccinium corymbosum*)	246–692	84–430	
Cranberry (*Vaccinium macrocarpon*)	120–315	20–66	
Bilberry (*Vaccinium myrtillus*)	181–585	299	
Blackcurrant (*Ribes nigrum*)	560	128–411	
Raspberry (*Rubus idaeus*)	126–2494	3–230	
Strawberries (*Fragaria ananassa*)	102–225	20–39	

GAE: Galic acid equivalents; C3GE: cyanidin-3-glucoside equivalents; FW: fresh weight.

1.2. Phenolic Acids

Phenolic acids are derived from benzoic and cinnamic acids and are synthesized through the phenylpropanoid metabolic pathway. They are commonly esterified with other molecules as carbohydrates and organic acids. Among the benzoic acid derivates in berries there are gallic acid, p-hydroxybenzoic acid, salicylic acid, and ellagic acid (Szajdek and Borowska, 2008; Manganaris et al., 2014). The most abundant in berries are caffeic acid, p-coumaric acid, and ferulic acid. It has been reported that large concentrations of these compounds are found in rowanberry (103 mg/100 g wet basis, FW), chokeberry (96 mg/100 g FW), and blueberry (85 mg/100g FW) (Mattila, Hellström, and Törrönen, 2006).

1.3. Flavonoids

Flavonoids are the most numerous and studied phenolic compounds in foods. Their basic structure consists of a diphenylpropane skeleton (flavan nucleus) formed by two external aromatic rings and a three-carbon bridge, which can be closed- (as flavones, flavanols, and anthocyanins) or open-chained (as chalcones). Their most important roles in plants include UV protection, pigmentation, stimulation of N_2 fixation nodes, and resistance against some diseases (Gwiazdon and Biesaga, 2017; Espino, Jiménez, and Cardador-Martínez, 2019). Flavonoids are classified into anthocyanins, flavanols, flavonols, flavones, flavanones, and isoflavonoids (Paredes-López et al., 2010).

Anthocyanins are especially abundant in berries, conferring them bright colors such as red, blue, and purple. They are synthesized in the cytosol and are released in the vacuole, where they are stored to be mostly found in the skin of fruits (Szajdek and Borowska, 2008; Jaakola, 2013). Anthocyanins are glucosides of anthocyanidins (Izabela and Wei, 2004) and structural differences between them mostly consist of the number of hydroxyl groups and position, type, and/or number of sugars linked to the molecule (Kong et al., 2003). Glycosylation in positions C-3 and C-5 of the molecule affect the color perception of anthocyanins. This is how derivates of 3-glycosides show a brighter color than do 3,5-diglycosides, while glucose acylation increases pigment stability even more (Flamini et al., 2013). Anthocyanin concentrations can vary between plant species and even between species of the same genus (Peña-Sanhueza et al., 2017). As shown in Table 2, blueberry exhibits a high content of monomeric anthocyanins (430 mg cyanidin-3-glucoside equivalents/100 g FW), closely followed by blackcurrant (411 mg cyanidin-3-glucoside equivalents/100 g FW).

Table 3 shows the profile of the anthocyanins identified in blackberry and raspberry (Marhuenda et al., 2016). In berries, glycosylated anthocyanins can be found in position C-3, which indeed favors their solubility and stability but compromises their ability to catch radicals. The sugars most commonly linked to anthocyanidins are glucose, galactose, rhamnose, arabinose, rutinose, and sambubiose. The six anthocyanidins predominant in fruits and berries are pelargonidin, cyanidin, delphinidin, peonidin, petunidin, and malvidin (Jaakola, 2013). The following anthocyanins have been identified in blackberry: cyanidin 3-glucoside, cyanidin 3-galactoside, cyanidin 3-xyloside, cyanidin 3-dioxalyl-glucoside, cyanidin 3-rutinoside, cyanidin 3-sophoroside, cyanidin 3-glucosylrutinoside, cyanidin 3-arabinoside, malvidin 3-arabinoside, pelargonidin 3-glucoside, cyanidin 3-(3-malonyl-glucoside), and cyanidin 3-(6-malonyl-glucoside) (Kaume, Howard, and Devareddy, 2012).

Table 3. Anthocyanin content (mg/100 g FW) in blackberry and raspberry

Anthocyanin	Blackberry	Raspberry
Cyanidin 3-glucoside	57.2–107	10.1–57.5
Cyanidin-3-sophoroside	41.2–62.0	ND
Petunidin 3-glucoside	ND	57.5
Cyanidin 3-glucosylrutinoside	ND	56.4–151
Cyanidin 3-rutinoside	0.35–25.0	ND–19.6
Cyanidin 3-xyloside	8.0–48.3	ND

ND: Non-detected. (Marhuenda et al., 2016; Schulz and Chim, 2019).

To promote the understanding of the absorption and metabolic processes of anthocyanins, these compounds have been divided into three groups: pelargonidin, cyanidin/peonidin, and multiple anthocyanins. It is well known that fruits rich in cyanidin and peonidin can metabolically convert to each other by methylation and demethylation. Furthermore, they can be metabolized in protocatechuic and vanillic acids. Berries as strawberry, lingonberry, and raspberry contain cyanidin as the predominant anthocyanin (Fang, 2015).

Flavonols as kaempferol, myricetin, and quercetin are the most common and can be found as *O*-glycosides. Over 50 flavonols have been identified in different species of wild and grown berries; the most predominant one is quercetin in most of the berries (Mikulic-Petkovsek et al., 2012). Quercetin-3-glucuronide is the most abundant glycosylated flavanol in strawberries (11 mg/kg FW), raspberries (1–6 mg/kg FW), and cloudberries (5 mg/kg FW) (Määttä-Riihinen, Kamal-Eldin, and Törrönen, 2004). The contribution of flavonols to the total antioxidant ability of berries is not higher than 14% when compared against that of anthocyanins (Manganaris et al., 2014). Flavonols as quercetin, kaempferol, and myricetin are attributed certain antimutagenic, anticancer, and antihypertensive biological activities (Espino, Jiménez, and Cardador-Martínez, 2019).

Flavanols found in fruits as monomeric chatequines (flavan-3-ols) are biosynthetic precursors of proanthocyanidins and are characterized by having a C6–C3–C6 skeleton with a hydroxyl group at position 3 of the ring C (Figure 1) (Colomer et al., 2016; Delcambre and Saucier, 2012). Chatequines have a trans configuration while epicatechins possess a cis configuration. Both compounds have two possible stereoisomers: (+)-catechin, (-)-catechin, (+)-epicatechin, and (-)-epicatechin. In grapes and wine, stereoisomers (+)-catechin, with a 2R,3S configuration (trans configuration), and (-)-epicatechin, with a 2R,3R configuration (cis configuration), are present. Additionally, it has been reported that (-)-epicatechin in grapes can be found esterified with gallic acid to create (-)-epicatechin gallate (Delcambre and Saucier, 2012). Schulz and Chim, (2019) reported 266–313 mg/100 g FW of catechin in blackberry, Sellappan, Akoh, and Krewer, (2002) reported 149–7300 mg/kg FW in raspberry.

1.4. Tannins

Tannins are commonly classified as hydrolyzable and condensed (proanthocyanidins). Hydrolyzable tannins are classified in gallotannins and ellagitannins. Gallotannins or galloyl glucoses are constituted by gallic acid and glucose esters, while hexahydroxydiphenic acid and glucose or quinic acid esters are the links for ellagitannins (Espino, Jiménez, and Cardador-Martínez, 2019). Condensed tannins or proanthocyanidins are oligomers and polymers from 2 to 200 flavan-3-ol monomers, with units generally constituted by catechins and epicatechins. Proanthocyanidins provide fruits with several properties, among which are astringency, bitterness, acidity, smell, and color (Rauf et al., 2019). It has been reported that there is a significant amount of proanthocyanidins in sea buckthorn pulp (1230.50

mg/100 mL), wild blueberry (1090.50 mg/100 mL), raspberry (981.67 mg/100 mL), and strawberry (860.41 mg/100 mL) (Hosseinian et al., 2007).

Ellagitannins are the second most important group of phenolic compounds in red raspberry. The main ellagitannin identified is sanguiin H-6, followed by lambertianin C. The sanguiin H-6 content in Glen Ample cultivar was 76 mg/100 g FW and that of lambertianin C, 31 mg/100 g FW (Burton-Freeman, Sandhu, and Edirisinghe, 2016).

1.5. Stilbenes

Stilbenes are formed by an essential structural skeleton consisting of two aromatic rings linked by an ethylene bridge (C6–C2–C6). Stilbenes exist in monomeric, oligomeric, and polymeric forms and they exhibit sugar, methyl, and methoxy substitutions along with other residues to form a great group of compounds called stilbenoids. The highest concentration of these compounds is found in the skin and seeds of berries. The stilbenes currently reported in berries are mostly E-resveratrol and E-piceid; their Z- and E-isomers are accumulated on the skin of berries during development stages (Blaszczyk, Sady, and Sielicka, 2019). The content of resveratrol reported in different edible berries is variable and has been reported as follows: bilberry 2.0–6.78 µg/g FW, blueberry 4 µg/g FW, cranberry 19.29 µg/g FW, and grape 1.13–121.62 µg/g FW (Može et al., 2011; Błaszczyk, Sady, and Sielicka, 2019). Stilbenes and their derivates have been studied in grapes; resveratrol has been particularly analyzed since it has been associated to the beneficial effects produced by wine. Approximately, 19 stilbene compounds have been found in grapes (Flamini et al., 2013).

1.6. Lignans

Lignans are formed by a carbon skeleton $(C6 - C3)_2$, which is the result of the chemical β,β' linkage of two phenylpropanes. They are classified in eight subgroups: arylnaphthalenes, aryltetralins, dibenzocyclooctadienes, dibenzylbutanes, dibenzylbutyrolactols, dibenzylbutyrolactones, furans, and furofurans (Rodríguez-García, Sánchez-Quesada, Toledo, Delgado-Rodríguez, & Gaforio, 2019). Recently, a study on lignan content was carried out in seeds and complete fruit of bilberries, cloudberries, blackberries, red gooseberries, and blackcurrants. The highest lignan concentration was reported in cloudberry, both in the seed (43,876 μg/100 g DW) and the whole fruit (14,775 μg/100 g DW), followed by blackberry (seed: 23,310 μg/100 g DW; whole fruit: 9,995 μg/100 g DW) and cranberry (seed: 16,713 μg/100 g DW; whole fruit: 6,858 μg/100 g DW). The predominant lignans in berries are lariciresinol-sesquilignan (LAR-SQ), secoisolariciresinol (SECO), and matairesinol (MR) (Smeds, Eklund, & Willför, 2012). In blackberries, SECO (3.72 mg/100 DW) and MR (<0.01 mg/100 g DW) lignans have been identified. When ingested by mammals, they are converted into compounds such as enterolactone and enterodiol by the colonic microbiota in the colon (Kaume, Howard and Devareddy, 2012).

2. Bioavailability of Phenolic Compounds

2.1. Absorption and Metabolism

According to the FDA, the term bioavailability refers to the "rate and extent to which the active ingredient or moiety is absorbed from a drug product and becomes available at the site of action." In general, the bioavailability of a nutrient or metabolite comprises gastrointestinal digestion, absorption, metabolism, distribution in tissues, and biological effect (Carbonell-Capella et al., 2014; Fang, 2014).

Several factors affect the bioavailability of phenolic compounds, such as their chemical structure (Talavera et al., 2004), pH alterations along the gastrointestinal tract (Stalmach et al.,2012), presence of other flavonoids (Walton et al., 2006), interaction with other diet components (as proteins, fats, and fibers) (Serafini et al., 2009), and the metabolism of the large intestine in charge of the gut microbiota (Jiménez-Girón et al., 2013).

The degree of absorption of anthocyanins, the most abundant flavonoids in *Rubus*, is influenced by the type of anthocyanidin, the glycosidic group, acylated groups, the dose administered, and stability against pH alterations (Fang, 2014). Glycosylated anthocyanins can be absorbed intact in their glycosylated form. Those that are not absorbed in the small intestine travel to the colon, where they are transformed into phenolic acids by the microorganisms in the small intestine. Afterwards, they are excreted in feces or absorbed into the mesenteric circulation (Kaume, Howard, and Devareddy, 2012).

Studies in animals have found that the systemic bioavailability of anthocyanins ranges between 0.26 and 1.8%. Studies in humans who consumed berries or grapes found maximum plasma levels of total anthocyanins in the interval from 1 to 100 nmol/L (Fang, 2014).

According to a follow-up study of phenolic compounds (anthocyanins, ellagic acid, and ellagitannins) in red raspberry, 40% of the ingested anthocyanins reach the large intestine, while 60% are absorbed, degraded, or remain undetected in the small intestine. Additionally, 23% of the ellagitannins ingested were found in the ileal liquid while a significant amount was hydrolyzed into ellagic acid (241% of the intake) in the large intestine (González-Barrio et al., 2010).

Phenolic compounds show phases I and II biotransformations in the human body. Phase I biotransformations consist of oxidation, reduction, and hydrolysis and occur less frequently (Kamiloglu et al., 2015). During the first stage of the reaction, changes in the structure of exogenous biomolecules take place. The aim is to improve the polarity of heterogeneous phenolic compounds and promote their excretion, which is carried out through the introduction of amino, carboxyl, and hydroxyl groups, among others (Hussain et al., 2019; Kamiloglu et al., 2015).

Phase II biotransformations take place in the liver and intestine and consist of conjugation reactions where several methyl, glucuronic, and sulfate derivates are formed (Cardona et al., 2013). It has been reported that the main metabolite of cyanidin-glucoside in humans is the protocatechuic acid and phloroglucinaldehyde, derived from rings A and B of cyanidin-3-glucoside (Cy-3-glc) (Figure 3) (Vitaglione et al., 2007; Kay, Kroon, and Cassidy, 2009). These metabolites can be further degraded into glucuronides and conjugated sulfates through the action of enzymes (Uridine 5′-diphosphate glucuronide transferase, sulfonate transferase, and catechol O-methyltransferase) (Kay, Kroon, and Cassidy, 2009; Hussain et al., 2019).

Another important reaction in Phase II of anthocyanins is methylation. Enzyme O-methyltransferase is involved in the O-methylation of flavonoids. Methylation improves their transport through biological membranes and efficiency in several biological activities as the antitumoral action in the Caco-2 cell line (Hussain et al., 2019; Kamiloglu et al., 2015).

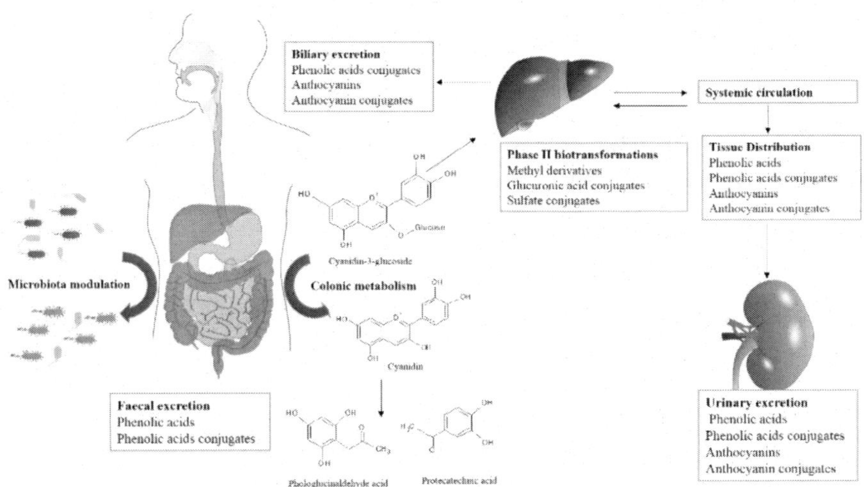

Figure 3. Schematic diagram of the metabolism of cyanidin 3-glucoside.

2.2. Interaction with Gut Microbiota

The term gut microbiota refers to the millions of microorganisms that inhabit mostly the large intestine and whose function is the fermentation of food components that escape digestion of the stomach and small intestine. These components can be carbohydrates, proteins, and fiber, producing short-chain fatty acids and vitamins K and B. In the colon, the microbial population is approximately 1012 cells/g of feces (Ley, Peterson, & Gordon, 2006). It is mainly constituted by seven bacterial divisions: *Firmicutes, Bacteroidetes, Actinobacteria, Fusobacteria, Proteobacteria, Verrucomicrobia,* and *Cyanobacteria.*

The interaction between the microbiota and phenolic compounds is bilateral (Figure 3). On one hand, gut microbiota can modulate the production, bioavailability, and biological activity of phenolic compounds. On the other hand, phenolic compounds have a prebiotic effect; that is, they can regulate the composition of the microbial population, disrupting the balance between *Bacteroidetes/Firmicutes* (Lee, Jenner, Low, & Lee, 2006).

The colonic microbiota is responsible for the degradation of phenolic compounds of high molecular weight into metabolites of lower weight. Lower molecular weight compounds as monomers or dimers are fast absorbed in the small intestine, but compounds with more complex structures, as oligomers and polymers can reach the colon without greater alterations (Monagas et al., 2010). Between 90 and 95% of the polyphenols ingested in the diet are accumulated in the colon, where they are degraded by the enzymatic action of the gut microbiota. In most cases, the resulting metabolites are biologically more active than the parent compounds (Selma, Espín, & Tomás-Barberán, 2009).

In general, the metabolism of the phenolic compounds in charge of the microbiota involves the rupture of the heterocyclic structure and the glycosidic bonds, producing lactones and aromatic and phenolic acids. The metabolism of anthocyanins involves the rupture of the heterocyclic ring C and the degradation of ring A, producing phloroglucinol derivates, while the degradation of ring B produces benzoic acids (Keppler & Humpf, 2005). Ellagitannins suffer hydrolysis in the intestinal lumen, producing free ellagic acid, which is metabolized by the gut microbiota and leads to urolithin generation. Two of the main genus involved in the metabolism of several phenolic compounds are *Clostridium* and *Eubacterium* (Selma, Espín, and Tomás-Barberán, 2009).

Metabolites of phenolic compounds produced by microbiota degradation can be absorbed or excreted in feces. When absorbed, they travel to the liver through the portal vein, where they undergo Phase II biotransformations, enter systemic circulation, and are distributed in tissues or excreted in urine.

Additionally, phenolic compounds can help to maintain a healthy gut microbiota since they stimulate the growth of beneficial bacteria and inhibit the growth of those that are pathogenic. For example, in an animal model of colon carcinogenesis researchers found that the group treated with red wine polyphenols showed elevated levels of *Bacteroides*, *Bifidobacterium*, and *Lactobacillus* spp; contrastingly, the control group mostly exhibited *Bacteroides*, *Clostridium*, and *Propionibacterium* spp (Dolara et al., 2005). The genera *Bifidobacterium* and *Lactobacillus* spp include bacteria that have proven to help the function of the intestinal barrier, modulate lipid metabolism, and stimulate the immune system of the host (Faria, Fernandes, Norberto, Mateus, & Calhau, 2014; Gibson, 2008). In contrast, bacteria from the genus *Clostridium* have been related to negative effects on health (Rastall et al., 2005).

2.3. Tissue Distribution

Concentrations of polyphenols and their accumulated metabolites in plasma and brain *in vivo* are low (high nanomolar, low micromolar) (Abd El Mohsen et al., 2002) when compared against other antioxidants as ascorbic acid and α-tocopherol (Halliwell, Zhao, & Whiteman, 2000). One of the most studied phenolic compounds present in *Rubus fruticosus* is Cy-3-glc, the main anthocyanin in blackberry (Cuevas-Rodríguez et al., 2010). In a study conducted in humans who ingested 500 mg Cy-3-glc, seventeen compounds were identified in serum and their most important degradation products were protocatechuic acid and phloroglucinaldehyde. The maximum concentration (C_{max}) of phenolic metabolites was identified within 10–2000 nM with t_{max} 2–30 h (De Ferrars et al., 2014). In another study in rats fed with a diet rich in anthocyanins (*Rubus fruticosus* L. extract) for 15 days, Talavera et al., (2005) observed that stomach tissue contained only native blackberry anthocyanins (Cy-3-glc and Cy-3-pentoside). Contrastingly, they identified native and methylated anthocyanins and conjugated anthocyanidins (cyanidin and peonidin monoglucuronides) in jejunum, liver, and kidney. The liver exhibited the largest proportion of methylated forms, while the jejunum and plasma also contained aglycone forms. Native blackberry anthocyanins and peonidin-3-O-glucoside (0.25 ± 0.05 nmol/g tissue) were found in the brain.

Polyphenol distribution in brain tissue has been scarcely studied; however, there are some studies in animals that prove polyphenols can cross the blood-brain barrier (BBB). Furthermore, they have been found in brain regions as hypothalamus, striatum, cortex, and hippocampus (Andres-lacueva et al., 2005; Kalt et al., 2008). The permeability of the BBB as polyphenols cross it is influenced by the chemical characteristics of the compound. The barrier is more permeable when less polar metabolites (O-methylated derivates) cross than when more polar compounds (sulfated and glucuronidated derivates) do. In addition, the interaction between polyphenols as specific transporters expressed in the BBB is another factor to consider (Youdim, Shukitt-Hale, & Joseph, 2004).

3. Neuroinflammation and Neurodegenerative Diseases

Nowadays, the incidence of neurodegenerative diseases as Alzheimer's (AD) and Parkinson's (PD) has increased. According to data reported by the World Health Organization, neurodegenerative diseases are expected to become the second cause of death worldwide by 2040, only behind cardiovascular disease and ahead of cancer (Gammon, 2014).

In brains of PD patients, the activity of complex I, an enzyme of the mitochondrial electron transport chain, has been found to be reduced, leading to the accumulation of reactive oxygen species (ROS) and promoting the formation of α-synuclein aggregates, which damage dopaminergic neurons in the substantia nigra (Blanchet et al., 2008; Hirsch, Hunot, and Harmann, 2005). Additionally, AD is the most common cause of dementia and has been identified as a metabolic disease caused by several factors, such as impaired insulin signaling and energy metabolism, oxidative stress, deposition of amyloid β peptide aggregates, and Tau protein hyperphosphorylation (Walsh et al., 2002; Grünblatt et al., 2007; La Monte, 2012).

However, neurodegenerative diseases in general are accompanied by neuroinflammation (Hirsch, Hunot, and Hartmann, 2005). It is a defense mechanism that protects the central nervous system (CNS) from lesions or infections. It is commonly a beneficial mechanism that helps to recover brain homeostasis after insult; still, an extended neuroinflammatory process can lead to neuronal damage (Spencer et al., 2012; Ransohoff, 2016).

In the CNS, microglial cells are in charge of the response against bacteria, viruses, and parasites. They have a similar action to that of peripheral macrophages (Graeber, Li, and Rodriguez, 2011). Glial activation favors the production of factors that contribute to neuronal damage (Figure 4), such as ROS, reactive nitrogen species (RNS), glutamate (Takeuchi et al., 2006), proinflammatory enzymes as inducible nitric oxide synthase (iNOS) and inducible cyclooxygenase-2 (COX-2), as well as proinflammatory cytokines like interleukin-1β (IL-1β) and tumor necrosis factor-α (TNF-α) (Lau, Bielinski, and Joseph, 2006; Shukitt-Hale et al., 2016).

Nevertheless, microglial activation by infections or lesions is not the only factor that triggers neurodegenerative diseases; aging is another important aspect to consider. As age increases, there is an evident decline in motor and cognitive functions (Pistell et al., 2011) directly related to a higher susceptibility among elderly individuals to long-term effects of oxidative stress and inflammation. The brain is highly susceptible to damage by oxidative stress since it has a deficient antioxidant defense system as a result of low concentrations of glutathione and moderate activity of endogenous antioxidant enzymes like catalase, superoxide dismutase, and glutathione peroxidase. In addition, it contains higher concentrations of transition metals as iron and copper, which contributes to ROS and RNS production (Aquilano et al., 2008).

Furthermore, recent in vivo studies reveal the key role that diet and intestinal microbiota play in the pathology of neurodegenerative diseases. The consumption of a high-fat diet (HFD) leads to an increase in oxidative stress, inflammation, loss of brain plasticity, behavioral deficits, and intestinal dysbiosis. The latter refers to an imbalance in the composition of gut microbiota and is involved in the development of metabolic disorders like obesity, metabolic syndrome, and type-2 diabetes (Baothman et al., 2016; Org et al., 2017). Intestinal dysbiosis can modify intestinal permeability, allowing access to potent proinflammatory endotoxins into the bloodstream, as lipopolysaccharide (LPS). LPS can trigger the peripheral inflammatory response and glial activation in the CNS, which leads to an increase in neuroinflammation (Qin et al., 2007).

4. Neuroprotective Effect of Polyphenols

Even though low concentrations of polyphenols or their metabolites have been found in the brain, these compounds have recently received particular attention because of their role in preventing and treating neurodegenerative disorders (Figure 4) (Fernández-Fernández et al., 2012; Hornedo Ortega et al., 2016). Their neuroprotective effects rely on their ability to cross the BBB and reach brain regions as the cortex and hippocampus, which are essential to memory and learning processes. Polyphenols can eliminate pathological concentrations of ROS and RNS, chelate metallic ions (Aquilano et al., 2008), and activate endogenous antioxidant enzymes, such as glutathione reductase, glutathione transferase, and superoxide dismutase (Gasparrini et al., 2017), thus preventing damage to proteins, lipids, and DNA.

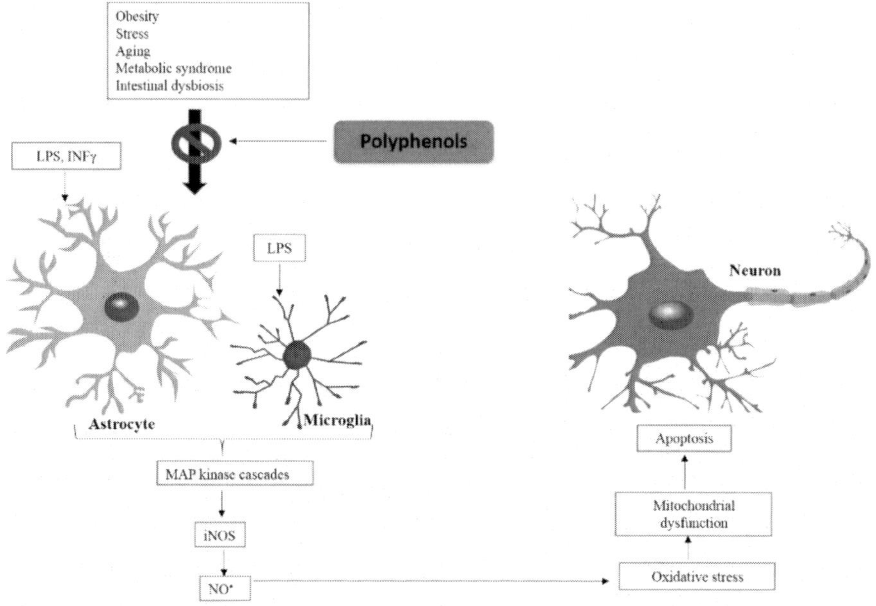

Figure 4. Role of polyphenols in microglial/astroglial activation and neurodegeneration. LPS: lipopolysaccharide, INFγ: interferon gamma, iNOS: inducible nitric oxide synthase, MAP kinases: mitogen activated protein kinases, NO: nitric oxide.

Several studies have demonstrated the anti-inflammatory and neuroprotective effect of the bioactive compounds present in blackberry (Table 4). Tavares et al., (2013) proved the neuroprotective effect of *Rubus brigantinus* and *Rubus vagabundus* since, during treatment, they observed a decrease in intracellular levels of ROS, modulation of glutathione levels, and activation of caspases in a neurodegeneration model. Later, Jung et al., (2015) reported that the intracellular levels of ROS decreased by 60% in RAW264.7 cells treated with a fraction rich in anthocyanins obtained from blackberry. Recently, Van de Velde et al., (2018) compared blackberry and strawberry extracts, finding that blackberry samples exhibited higher ROS reduction in RAW264.7 macrophages stimulated with LPS than strawberry samples did. Contrastingly, strawberry extracts were more active against IL-β and IL-6 gene expressions than the similar fractions from blackberry. They also proved that the extract of *Rubus fruticosus* rich in proanthocyanidins was the most active polyphenol fraction against iNOS gene expression (65%), suggesting that proanthocyanidins can be more active than anthocyanins against oxidative stress.

Furthermore, studies in animals fed with berries have confirmed the potential of phenolic compounds to improve cognitive functions and reverse damage to memory and learning caused by aging (Andres-lacueva et al., 2005; Williams et al., 2008; Malin et al., 2011). That beneficial effect arises from their ability to mitigate inflammation and oxidative stress in microglial cells (Poulose et al., 2012). Polyphenols can also modulate intracellular signaling pathways, as the nuclear factor -kβ (NF-kβ) cascade and mitogen-activated protein kinase (MAPK) pathway (Spencer et al., 2012), and regulate apoptosis processes and mitochondrial function (Spencer, Vauzour, and Rendeiro, 2009). They can also stimulate neuronal regeneration and induce neurogenesis.

A study with aged rats (19 months) fed with a diet supplemented with blackberry (2%) for 3 weeks demonstrated that the animals in the treated group showed a better motor performance than controls in three tests related to balance and coordination: accelerating rotarod, wire suspension, and small-plank walk. The results of the Morris water maze demonstrated that rats fed with blackberry showed a significantly higher performance in the short-term memory test (Shukitt-Hale, Cheng, and Joseph, 2009).

In Parkinson Disease models, extracts rich in anthocyanins and proanthocyanidins can improve mitochondrial function (Strathearn et al., 2014), while red wine polyphenols (resveratrol) have proven to protect nigral neurons from the damage caused by the administration of neurotoxin 1-methyl-4-phenyl-1,2,3,6-tetrahydropyridine (MPTP) (Blanchet et al., 2008). On the other hand, in animal models of AD, polyphenols are capable of modulating metabolism and aggregation of Aβ peptide (Wang et al., 2014; Wang et al., 2015), which plays a key role in the development of the disease.

Table 4. Neuroprotective effects of phenolic compounds of berries

Natural compounds	Animal models	Doses	Therapeutic effect	References
Blackberry	Aged Rats (19 months)	2% blackberry supplemented diet for 8 weeks	Improving Morris water maze performance and performance in motor test	(Shukitt-Hale, Cheng, and Joseph 2009)
Blackberry (*Rubus fruticosus)*	Rats fed a high fat diet	25 mg kg^{-1} body weight for 17 weeks	Modulating fractalkine and the thymus chemokine TCK-1 expression in hippocampus. In cortex decreasing the expression of CINC-3, CNTF, PDGF-AA, IL-10, TIMP-1 and RAGE	(Meireles et al., 2015)
Blackberry (*Rubus fruticosus)*	Rats fed a high fat diet	25 mg kg^{-1} body weight for 17 weeks	Modulating gut microbiota composition Increasing the production of kynurenic acid	(Marques et al., 2018)
Red raspberry (*Rubus iadeaus*)	Old mice fed a high fat diet	4% freeze-dried red raspberry	Improving novel object recognition and Barnes maze performance Decreasing IL-6 levels Increasing BDNF levels	(Carey et al., 2019)
Blueberry	4-month old male rats	2% blueberry diet for 8 weeks before kainic acid injection	Mitigating learning impairments following neurotoxic insult Anti-inflammatory activity	(Shukitt-Hale et al., 2016)
Blueberry (*Vaccinium angustifolium* Aiton)	Older adults	Blueberry juice (6-9 mL/kg body weight) for 12 weeks	Improving memory function	(Krikorian et al., 2010)

Abbreviations. CINC-3: cytokine-induced neutrophil chemoattractant, CNTF: ciliary neurotrophic factor, PDGF-AA: platelet-derived growth factor, IL-10: interleucina 10, TIMP-1: tissue inhibitor ofmetalloproteinase, RAGE: receptor for advanced glycation end products, IL-6: interleukin- 6, BDNF: brain-derived neurotrophic factor.

The chronic consumption of *Rubus fruticosus* extract rich in anthocyanins was able to prevent the damaging effect of neuroinflammation caused by HFD (Table 4) (Meireles et al., 2015). In another study conducted by the same research group, Wistar rats given HFD showed significant changes in the composition of the microbiota (dysbiosis), which could trigger neuroinflammation. However, in those animals administered HFD and blackberry extract for 17 weeks, the researchers observed that anthocyanins can modulate the composition of the gut microbiota and counter neuroinflammation. They identified elevated levels of kynurenic acid, a tryptophan metabolite that can act as a potent neuroprotector (Marques et al., 2018).

Polyphenols can also modulate the synthesis of neurotransmitters like acetylcholine and neurotrophic factors as brain-derived neurotrophic factor (BDNF). An increase in BDNF levels has been correlated with the formation of short and long- term memory and this in turn with cognitive improvements. Carey et al. (2019) studied the effect of raspberry (*Rubus idaeus*) consumption in old mice that received HFD. The HFD group showed damage in the novel object recognition test and elevated levels of IL-6 in the brain cortex. Contrastingly, the group fed with HDF supplemented with 4% red raspberry showed a better performance in the Barnes maze and the novel object recognition test, along with reduced IL-6 levels and increased BDNF concentrations.

CONCLUSION

This chapter summarizes the available scientific information of nutritional value, bioactive compound content, bioavailability (colonic interaction with gut microbiota), and the neuroprotective effect of the *Rubus* genus. There is no doubt that the *Rubus* genus group plant species are a great source of nutrients and nutraceutical compounds; because of the increasing demand of consumers, their intake has shown a rising trend. They constitute a relevant source of bioactive substances, as phenolic compounds, vitamins, and carotenoids. These phenolic compounds can offer anti-inflammatory/antioxidant capacity and neuroprotective effects. In general, the metabolites produced by the enzymatic action of the intestinal microbiota from ingested phenolic compounds show a greater biological action than the parent compounds. They are able to penetrate the Blood-Brain Barrier and reach the brain in low concentrations sufficient to modulate signaling pathways involved in the formation of short- and long-term memory. In addition, these compounds have also been shown to attenuate microglia/astrocyte-mediated neuroinflammation. For all the above, nutritional interventions with foods rich in phenolic compounds are an alternative for the prevention of neurodegenerative diseases. Still, one of the challenges for future research is to obtain further results from human trials to completely elucidate the action mechanism of phenolic compounds.

REFERENCES

Abd El Mohsen, Manal M., Gunter Kuhnle, Andreas R. Rechner, Hagen Schroeter, Sarah Rose, Peter Jenner, and Catherine A. Rice-Evans. 2002. “Uptake and Metabolism of Epicatechin and Its Access to the Brain after Oral Ingestion.” *Free Radical Biology and Medicine* 33 (12): 1693–1702. https://doi.org/10.1016/S0891-5849(02)01137-1.

Ahmad, Mushtaq, Saima Masood, Shazia Sultana, Taibi Ben Hadda, Ammar Bader, and Muhammad Zafar. 2015. “Antioxidant and Nutraceutical Value of Wild Medicinal Rubus Berries.” *Pakistan Journal of Pharmaceutical Sciences* 28 (1): 241–47.

Andres-lacueva, Cristina, Barbara Shukitt-hale, Rachel L Galli, Olga Jauregui, Rosa M Lamuela-raventos, and James Joseph. 2005. “Anthocyanins in Aged Blueberry-Fed Rats Are Found Centrally and May Enhance Memory.” *Nutritional Neuroscience* 8 (2): 111–20. https://doi.org/10.1080/10284150500078117.

Aquilano, Katia, Sara Baldelli, Giuseppe Rotilio, and Maria Rosa Ciriolo. 2008. “Role of Nitric Oxide Synthases in Parkinson’s Disease: A Review on the Antioxidant and Anti-Inflammatory Activity of

Polyphenols." *Neurochemical Research* 33 (12): 2416–26. https://doi.org/10.1007/s11064-008-9697-6.

Baothman, Othman A., Mazin A. Zamzami, Ibrahim Taher, Jehad Abubaker, and Mohamed Abu-Farha. 2016. "The Role of Gut Microbiota in the Development of Obesity and Diabetes." *Lipids in Health and Disease* 15 (1): 1–8. https://doi.org/10.1186/s12944-016-0278-4.

Blanchet, Julie, Fanny Longpré, Geneviève Bureau, Marc Morissette, Thérèse DiPaolo, Gilles Bronchti, and Maria Grazia Martinoli. 2008. "Resveratrol, a Red Wine Polyphenol, Protects Dopaminergic Neurons in MPTP-Treated Mice." *Progress in Neuro-Psychopharmacology and Biological Psychiatry* 32 (5): 1243–50. https://doi.org/10.1016/j.pnpbp.2008.03.024.

Blaszczyk, Alfred, Sylwia Sady, and Maria Sielicka. 2019. "The Stilbene Profile in Edible Berries." *Phytochemistry Reviews* 18 (1): 37–67. https://doi.org/10.1007/s11101-018-9580-2.

Burton-Freeman, Britt M, Amandeep K Sandhu, and Indika Edirisinghe. 2016. "Red Raspberries and Their Bioactive Polyphenols: Cardiometabolic and Neuronal Health Links." *Advances in Nutrition* 7 (1): 44–65. https://doi.org/10.3945/an.115.009639.

Carbonell-Capella, Juana M., Magdalena Buniowska, Francisco J. Barba, María J. Esteve, and Ana Frígola. 2014. "Analytical Methods for Determining Bioavailability and Bioaccessibility of Bioactive Compounds from Fruits and Vegetables: A Review." *Comprehensive Reviews in Food Science and Food Safety* 13 (2): 155–71. https://doi.org/10.1111/1541-4337.12049.

Cardona, Fernando, Cristina Andrés-Lacueva, Sara Tulipani, Francisco J. Tinahones, and María Isabel Queipo-Ortuño. 2013. "Benefits of Polyphenols on Gut Microbiota and Implications in Human Health." *Journal of Nutritional Biochemistry* 24 (8): 1415–22. https://doi.org/10.1016/j.jnutbio.2013.05.001.

Carey, Amanda N., Giulia I. Pintea, Shelby Van Leuven, Kelsea R. Gildawie, Laura Squiccimara, Elizabeth Fine, Abigail Rovnak, and Mark Harrington. 2019. "Red Raspberry (Rubus Ideaus)

Supplementation Mitigates the Effects of a High-Fat Diet on Brain and Behavior in Mice." *Nutritional Neuroscience* 0 (0): 1–11. https://doi.org/10.1080/1028415x.2019.1641284.

Colomer, Ramon, Ariadna Sarrats, Ruth Lupu, and Teresa Puig. 2016. "Natural Polyphenols and Their Synthetic Analogs as Emerging Anticancer Agents." *Current Drug Targets* 18 (2): 147–59. https://doi.org/10.2174/1389450117666160112113930.

Crozier, Alan, Takao Yokota, Indu B. Jaganath, Serena Marks, Michael Saltmarsh, and Michael N. Clifford. 2006. "Secondary Metabolites in Fruits, Vegetables, Beverages and Other Plant-Based Dietary Components." Edited by Alan Crozier, Michael N. Clifford, and Hiroshi Ashihara. *Plant Secondary Metabolites: Occurrence, Structure and Role in the Human Diet*, 208–302. https://doi.org/10.1002/9780470988558.ch2.

Cuevas-Rodríguez, Edith O., Gad G. Yousef, Pedro A. García-Saucedo, José López-Medina, Octavio Paredes-López, and Mary Ann Lila. 2010. "Characterization of Anthocyanins and Proanthocyanidins in Wild and Domesticated Mexican Blackberries (Rubus Spp.)." *Journal of Agricultural and Food Chemistry* 58 (12): 7458–64. https://doi.org/10.1021/jf101485r.

Delcambre, Adéline, and Cédric Saucier. 2012. "Identification of New Flavan-3-Ol Monoglycosides by UHPLC-ESI-Q-TOF in Grapes and Wine." *Journal of Mass Spectrometry* 47 (6): 727–36. https://doi.org/10.1002/jms.3007.

Dolara, Piero, Cristina Luceri, Carlotta De Filippo, Angelo Pietro Femia, Lisa Giovannelli, Giovanna Caderni, Cinzia Cecchini, Stefania Silvi, Carla Orpianesi, and Alberto Cresci. 2005. "Red Wine Polyphenols Influence Carcinogenesis, Intestinal Microflora, Oxidative Damage and Gene Expression Profiles of Colonic Mucosa in F344 Rats." *Mutation Research - Fundamental and Molecular Mechanisms of Mutagenesis* 591 (1–2): 237–46. https://doi.org/10.1016/j.mrfmmm.2005.04.022.

Elisha, Ishaku Leo, Jean Paul Dzoyem, Lyndy Joy McGaw, Francien S. Botha, and Jacobus Nicolaas Eloff. 2016. "The Anti-Arthritic, Anti-Inflammatory, Antioxidant Activity and Relationships with Total

Phenolics and Total Flavonoids of Nine South African Plants Used Traditionally to Treat Arthritis." *BMC Complementary and Alternative Medicine* 16 (1): 1–10. https://doi.org/10.1186/s12906-016-1301-z.

Espino, S. María Teresa, Martínez Cristian Jiménez, and Anaberta Cardador-Martínez. 2019. "Phenolic Compounds in Cereals and Legumes." Edited by Leo M.L. Nollet and Jane A. Gutierrez-Uribe. *Phenolic Compounds in Food*, 319–32. https://doi.org/10.1201/9781315120157-16.

Fang, Jim. 2014. "*Bioavailability of Anthocyanins*" 2532 (4): 508–20. https://doi.org/10.3109/03602532.2014.978080.

Fang, Jim. 2015. "Classification of Fruits Based on Anthocyanin Types and Relevance to Their Health Effects." *Nutrition* 31 (11–12): 1301–6. https://doi.org/10.1016/j.nut.2015.04.015.This.

Faria, Ana., Iva. Fernandes, Sónia. Norberto, Nuno. Mateus, and Conceição. Calhau. 2014. "Interplay between Anthocyanins and Gut Microbiota." *Journal of Agricultural and Food Chemistry* 62 (29): 6898–6902. https://doi.org/10.1021/jf501808a.

Fernández-Fernández, Laura, Gemma Comes, Irene Bolea, Tony Valente, Jessica Ruiz, Patricia Murtra, Bartolomé Ramirez, et al., 2012. "LMN Diet, Rich in Polyphenols and Polyunsaturated Fatty Acids, Improves Mouse Cognitive Decline Associated with Aging and Alzheimer ' s Disease." *Behavioural Brain Research* 228 (2): 261–71. https://doi.org/10.1016/j.bbr.2011.11.014.

Ferrars, R. M. De, C. Czank, Q. Zhang, N. P. Botting, P. A. Kroon, A. Cassidy, and C. D. Kay. 2014. "The Pharmacokinetics of Anthocyanins and Their Metabolites in Humans." *British Journal of Pharmacology* 171 (13): 3268–82. https://doi.org/10.1111/bph.12676.

Flamini, Riccardo, Fulvio Mattivi, Mirko De Rosso, Panagiotis Arapitsas, and Luigi Bavaresco. 2013. "Advanced Knowledge of Three Important Classes of Grape Phenolics: Anthocyanins, Stilbenes and Flavonols." *International Journal of Molecular Sciences* 14 (10): 19651–69. https://doi.org/10.3390/ijms141019651.

Gammon, Katharine. 2014. "Brain Windfall." *Nature Carreers* 515: 299–300.

Gasparrini, Massimiliano, Tamara Y. Forbes-Hernandez, Francesca Giampieri, Sadia Afrin, Josè M. Alvarez-Suarez, Luca Mazzoni, Bruno Mezzetti, Josè L. Quiles, and Maurizio Battino. 2017. "Anti-Inflammatory Effect of Strawberry Extract against LPS-Induced Stress in RAW 264.7 Macrophages." *Food and Chemical Toxicology* 102: 1–10. https://doi.org/10.1016/j.fct.2017.01.018.

Gibson, Glenn R. 2008. "Prebiotics as Gut Microflora Management Tools." *Journal of Clinical Gastroenterology* 42 Suppl 2 (July): 75–79. https://doi.org/10.1097/mcg.0b013e31815ed097.

González-Barrio, Rocío, Gina Borges, William Mullen, and Alan Crozier. 2010. "Bioavailability of Anthocyanins and Ellagitannins Following Consumption of Raspberries by Healthy Humans and Subjects with an Ileostomy." *Journal of Agricultural and Food Chemistry* 58 (7): 3933–39. https://doi.org/10.1021/jf100315d.

Graeber, Manuel B., Wei Li, and Michael L. Rodriguez. 2011. "Role of Microglia in CNS Inflammation." *FEBS Letters* 585 (23): 3798–3805. https://doi.org/10.1016/j.febslet.2011.08.033.

Grünblatt, Edna, Melita Salkovic-Petrisic, Jelena Osmanovic, Peter Riederer, and Siegfried Hoyer. 2007. "Brain Insulin System Dysfunction in Streptozotocin Intracerebroventricularly Treated Rats Generates Hyperphosphorylated Tau Protein." *Journal of Neurochemistry* 101 (3): 757–70. https://doi.org/10.1111/j.1471-4159.2006.04368.x.

Gwiazdon, Malgorzata, and Magdalena Biesaga. 2017. "Phenolic Compounds in Plant Materials: Problems and New Analytical Solutions." In *Phenolic Compounds Types, Effects and Research*, edited by Teresa Garde-Cerdán, Ana Gonzalo-Diago, and Eva Pérez-Álvarez, 1–27. New Jersey, USA: Nova Science Publishers, Inc.

Halliwell, Barry, Kaicun Zhao, and Matthew Whiteman. 2000. "The Gastrointestinal Tract: A Major Site of Antioxidant Action" *Free Radical Research* 33 (6): 819–30. https://doi.org/10.1080/10715760000301341.

Hariram, Shivraj, Nile, Se Won, Park. 2014. "Edible Berries : Review on Bioactive Components and Their Effect on Human Health." *Nutrition* 30 (2): 134–44.

Hirsch, Etienne C., Stéphane Hunot, and Andreas Hartmann. 2005. "Neuroinflammatory Processes in Parkinson's Disease." *Parkinsonism and Related Disorders* 11 (SUPPL. 1): 9–15. https://doi.org/10.1016/j.parkreldis.2004.10.013.

Hornedo Ortega, Ruth, Alvarez Fernández María Antonia, Ana Belén Cerezo, Tristan Richard, Ana María Troncoso, Carmen Garcia-Parrilla, and Ana Mar. 2016. "Protocatechuic Acid: Inhibition of Fibril Formation, Destabilization of Preformed Fibrils of Amyloid - β and α - Synuclein, and Neuroprotection." https://doi.org/10.1021/acs.jafc.6b03217.

Hosseinian, Farah S., Wende Li, Arnie W. Hydamaka, Apollinaire Tsopmo, Lynda Lowry, James Friel, and Trust Beta. 2007. "Proanthocyanidin Profile and ORAC Values of Manitoba Berries, Chokecherries, and Seabuckthorn." *Journal of Agricultural and Food Chemistry* 55 (17): 6970–76. https://doi.org/10.1021/jf071163a.

Hummer, K. E., 1996. Rubus diversity. *Hort Science*. 31:182–183

Hummer, K. E. 2010. Rubus Pharmacology: Antiquity to the Present. *HortScience* 45(11): 1587-1591. https://doi.org/10.21273/HORTSCI.45.11.1587.

Hussain, Muhammad Bilal, Sadia Hassan, Marwa Waheed, Ahsan Javed, Muhammad Adil Farooq, and Ali Tahir. 2019. "Bioavailability and Metabolic Pathway of Phenolic Compounds." *Plant Physiological Aspects of Phenolic Compounds*. 2019. https://doi.org/http://dx.doi.org/10.5772/57353.

Izabela, Konczak, and Zhang Wei. 2004. "Anthocyanins-More than Nature' s Colours." *Journal of Biomedicine and Biotechnology* 5: 239–40.

Jaakola, Laura. 2013. "New Insights into the Regulation of Anthocyanin Biosynthesis in Fruits." *Trends in Plant Science* 18 (9): 477–83. https://doi.org/10.1016/j.tplants.2013.06.003.

Jiménez-Girón, A., I. Muñoz-González, M. Monagas, P. J. Martín, M. Mora, F. J. Tinahones, C. Andrés-Lacueva, and B. Bartolomé. 2013.

"Comparative Study of Microbial-Derived Phenolic Metabolites In." *Journal of Agricultural and Food Chemistry* 61: 3909–15.

Jung, Hana, Hee Jae Lee, Hyunnho Cho, Kiuk Lee, Ho Kyung Kwak, and Keum Taek Hwang. 2015. "Anthocyanins in Rubus Fruits and Antioxidant and Anti-Inflammatory Activities in RAW 264.7 Cells." *Food Science and Biotechnology* 24 (5): 1879–86. https://doi.org/10.1007/s10068-015-0246-1.

Kalt, Wilhelmina, Jeffrey B. Blumberg, Jane E. McDonald, Melinda R. Vinqvist-Tymchuk, Sherry A.E. Fillmore, Brigitte A. Graf, Jennifer M. O'leary, and Paul E. Milbury. 2008. "Identification of Anthocyanins in the Liver, Eye, and Brain of Blueberry-Fed Pigs." *Journal of Agricultural and Food Chemistry* 56 (3): 705–12. https://doi.org/10.1021/jf071998l.

Kamiloglu, Senem, Esra Capanoglu, Charlotte Grootaert, and John van Camp. 2015. "Anthocyanin Absorption and Metabolism by Human Intestinal Caco-2 Cells—A Review." *International Journal of Molecular Sciences* 16 (9): 21555–74. https://doi.org/10.3390/ijms160921555.

Kaume, Lydia, Luke R. Howard, and Latha Devareddy. 2012. "The Blackberry Fruit: A Review on Its Composition and Chemistry, Metabolism and Bioavailability, and Health Benefits." *Journal of Agricultural and Food Chemistry* 60 (23): 5716–27. https://doi.org/10.1021/jf203318p.

Kay, Colin D., Paul A. Kroon, and Aedin Cassidy. 2009. "The Bioactivity of Dietary Anthocyanins Is Likely to Be Mediated by Their Degradation Products." *Molecular Nutrition and Food Research* 53 (SUPPL. 1): 92–101. https://doi.org/10.1002/mnfr.200800461.

Kelsey, Natalie, Whitney Hulick, Aimee Winter, Erika Ross, and Daniel Linseman. 2011. "*Neuroprotective Effects of Anthocyanins on Apoptosis Induced by Mitochondrial Oxidative Stress*" 14 (6): 249–60.

Keppler, Katrin, and Hans-ulrich Humpf. 2005. "Metabolism of Anthocyanins and Their Phenolic Degradation Products by the Intestinal Microflora" 13: 5195–5205. https://doi.org/10.1016/j.bmc.2005.05.003.

Kong, Jin Ming, Lian Sai Chia, Ngoh Khang Goh, Tet Fatt Chia, and R. Brouillard. 2003. "Analysis and Biological Activities of Anthocyanins." *Phytochemistry* 64 (5): 923–33. https://doi.org/10.1016/S0031-9422(03)00438-2.

La Monte, Suzanne M. De. 2012. "Contributions of Brain Insulin Resistance and Deficiency in Amyloid-Related Neurodegeneration in Alzheimers Disease." *Drugs* 72 (1): 49–66. https://doi.org/10.2165/11597760-000000000-00000.

Lau, Francis C., Donna F Bielinski, and James A. Joseph. 2006. "*Inhibitory Effects of Blueberry Extract on the Production of Inflammatory Mediators in Lipopolysaccharide-Activated BV2 Microglia*" 693 (December 2005): 680–93. https://doi.org/10.1002/jnr.

Ley, Ruth E., Daniel A. Peterson, and Jeffrey I. Gordon. 2006. "Ecological and Evolutionary Forces Shaping Microbial Diversity in the Human Intestine." *Cell* 124 (4): 837–48. https://doi.org/10.1016/j.cell.2006.02.017.

Lee, Hui Cheng, Andrew M. Jenner, Chin Seng Low, and Yuan Kun Lee. 2006. "Effect of Tea Phenolics and Their Aromatic Fecal Bacterial Metabolites on Intestinal Microbiota." *Research in Microbiology* 157 (9): 876–84. https://doi.org/10.1016/j.resmic.2006.07.004.

Määttä-Riihinen, Kaisu R., Afaf Kamal-Eldin, and A. Riitta Törrönen. 2004. "Identification and Quantification of Phenolic Compounds in Berries of Fragaria and Rubus Species (Family Rosaceae)." *Journal of Agricultural and Food Chemistry* 52 (20): 6178–87. https://doi.org/10.1021/jf049450r.

Malin, David H, D Ph, David R Lee M A, M S, Pilar Goyarzu, D Ph, Yu-hsuan Chang M A, et al., 2011. "Short-Term Blueberry-Enriched Diet Prevents and Reverses Object Recognition Memory Loss in Aging Rats." *Nutrition* 27 (3): 338–42. https://doi.org/10.1016/j.nut.2010.05.001.

Manganaris, George A, Vlasios Goulas, R Vicente, and Leon A Terry. 2014. "Berry Antioxidants : Small Fruits Providing Large Benefits." *Journal of the Science of Food and Agriculture* 94: 825–33. https://doi.org/10.1002/jsfa.6432.

Marhuenda, Javier, María Dolores Alemán, Amadeo Gironés-Vilaplana, Alfonso Pérez, Gabriel Caravaca, Fernando Figueroa, Juana Mulero, and Pilar Zafrilla. 2016. "Phenolic Composition, Antioxidant Activity, and in Vitro Availability of Four Different Berries." *Journal of Chemistry* 2016. https://doi.org/10.1155/2016/5194901.

Marques, Cláudia, Iva Fernandes, Manuela Meireles, Ana Faria, Jeremy P.E. Spencer, Nuno Mateus, and Conceição Calhau. 2018. "Gut Microbiota Modulation Accounts for the Neuroprotective Properties of Anthocyanins." *Scientific Reports* 8 (1): 1–9. https://doi.org/10.1038/s41598-018-29744-5.

Mattila, Pirjo, Jarkko Hellström, and Riitta Törrönen. 2006. "Phenolic Acids in Berries, Fruits, and Beverages." *Journal of Agricultural and Food Chemistry* 54 (19): 7193–99. https://doi.org/10.1021/jf0615247.

Meireles, Manuela, Claudia Marques, Sonia Norberto, Iva Fernandes, Nuno Mateus, Catarina Rendeiro, Jeremy P E Spencer, Ana Faria, and, Conceição Calhau. 2015. "The Impact of Chronic Blackberry Intake on the Neuroinflammatory Status of Rats Fed a Standard or High-Fat Diet." *Journal of Nutritional Biochemistry* 26 (11): 1166–73. https://doi.org/10.1016/j.jnutbio.2015.05.008.

Mikulic-Petkovsek, Maja, Ana Slatnar, Franci Stampar, and Robert Veberic. 2012. "HPLC-MS n Identification and Quantification of Flavonol Glycosides in 28 Wild and Cultivated Berry Species." *Food Chemistry* 135 (4): 2138–46. https://doi.org/10.1016/j.foodchem.2012.06.115.

Monagas, Maria, Mireia Urpi-Sarda, Fernando Sánchez-Patán, Rafael Llorach, Ignacio Garrido, Carmen Gómez-Cordovés, Cristina Andres-Lacueva, and Begoña Bartolomé. 2010. "Insights into the Metabolism and Microbial Biotransformation of Dietary Flavan-3-Ols and the Bioactivity of Their Metabolites." *Food and Function* 1 (3): 233–53. https://doi.org/10.1039/c0fo00132e.

Može, Špela, Tomaž Polak, Lea Gašperlin, Darinka Koron, Andreja Vanzo, Nataša Poklar Ulrih, and Veronika Abram. 2011. "Phenolics in Slovenian Bilberries (Vaccinium Myrtillus L.) and Blueberries (Vaccinium Corymbosum L.)." *Journal of Agricultural and Food Chemistry* 59 (13): 6998–7004. https://doi.org/10.1021/jf200765n.

Olas, Beata. 2016. "Sea Buckthorn as a Source of Important Bioactive Compounds in Cardiovascular Diseases." *Food and Chemical Toxicology* 97: 199–204. https://doi.org/10.1016/j.fct.2016.09.008.

Olas, Beata. 2018. "Berry Phenolic Antioxidants – Implications for Human Health?" *Frontiers in Pharmacology* 9 (78): 1–14. https://doi.org/10.3389/fphar.2018.00078.

Org, Elin, Yuna Blum, Silva Kasela, Margarete Mehrabian, Johanna Kuusisto, Antti J. Kangas, Pasi Soininen, et al., 2017. "Relationships between Gut Microbiota, Plasma Metabolites, and Metabolic Syndrome Traits in the METSIM Cohort." *Genome Biology* 18 (1): 1–14. https://doi.org/10.1186/s13059-017-1194-2.

Paredes-López, Octavio, Martha L. Cervantes-Ceja, Mónica Vigna-Pérez, and Talía Hernández-Pérez. 2010. "Berries: Improving Human Health and Healthy Aging, and Promoting Quality Life-A Review." *Plant Foods for Human Nutrition* 65: 299–308. https://doi.org/10.1007/s11130-010-0177-1.

Peña-Sanhueza, Daniela, Claudio Inostroza-Blancheteau, Alejandra Ribera-Fonseca, and Marjorie Reyes-Díaz. 2017. "Anthocyanins in Berries and Their Potential Use in Human Health." *Superfood and Functional Food - The Development of Superfoods and Their Roles as Medicine*. InTech. 2017. https://doi.org/10.5772/67104.

Pistell, Paul J., Edward L. Spangler, Bennett Kelly-Bell, Marshall G. Miller, Rafael. de Cabo, and Donald K. Ingram. 2011. "Age-Associated Learning and Memory Deficits in Two Mouse Versions of the Stone T-Maze." *Bone* 23 (1): 1–7. https://doi.org/10.1038/jid.2014.371.

Poulose, Shibu M, Derek R Fisher, Jessica Larson, Donna F Bielinski, Agnes M Rimando, Amanda N Carey, Alexander G Schauss, and Barbara Shukitt-hale. 2012. "*Anthocyanin-Rich Ac a¸ i (Euterpe Oleracea Mart.) Fruit Pulp Fractions Attenuate Inflammatory Stress Signaling in Mouse Brain BV-2 Microglial Cells*."

Qin, Liya, Xuefei Wu, Michelle Block, Yuxin Liu, George Breese, Jau-shyong Hong, Darin Knapp, and Fulton Crews. 2007. "Systemic LPS Causes Chronic Neuroinflammation and Progressive

Neurodegeneration." *Glia* 55 (12): 1251–62. https://doi.org/10.1002/glia.

Ransohoff, Richard M. 2016. "*How Neuroinflammation Contributes to Neurodegeneration*" 353 (6301): 168–75.

Rastall, Robert A., Glenn R. Gibson, Harsharnjit S. Gill, Fransisco Guarner, Todd R. Klaenhammer, Bruno Pot, Gregor Reid, Ian R. Rowland, and Mary Ellen Sanders. 2005. "Modulation of the Microbial Ecology of the Human Colon by Probiotics, Prebiotics and Synbiotics to Enhance Human Health: An Overview of Enabling Science and Potential Applications." *FEMS Microbiology Ecology* 52 (2): 145–52. https://doi.org/10.1016/j.femsec.2005.01.003.

Rauf, Abdur, Muhammad Imran, Tareq Abu-izneid, and Seema Patel. 2019. "Proanthocyanidins: A Comprehensive Review." *Biomedicine & Pharmacotherapy* 116: 108999. https://doi.org/10.1016/j.biopha.2019.108999.

Rodríguez-García, Carmen, Cristina Sánchez-Quesada, Estefanía Toledo, Miguel Delgado-Rodríguez, and José J. Gaforio. 2019. "Naturally Lignan-Rich Foods: A Dietary Tool for Health Promotion?" *Molecules* 24 (917): 1–25. https://doi.org/10.3390/molecules24050917.

Schulz, Mayara, and Josiane Freitas Chim. 2019. "Nutritional and Bioactive Value of Rubus Berries." *Food Bioscience* 31: 100438. https://doi.org/10.1016/j.fbio.2019.100438.

Seeram, Navindra P., Lynn S. Adams, Yanjun Zhang, Rupo Lee, Daniel Sand, Henry S. Scheuller, and David Heber. 2006. "Blackberry, Black Raspberry, Blueberry, Cranberry, Red Raspberry , and Strawberry Extracts Inhibit Growth and Stimulate Apoptosis of Human Cancer Cells In Vitro." *Journal of Agricultural and Food Chemistry* 54: 9329–39.

Sellappan, Subramani, Casimir C. Akoh, and Gerard Krewer. 2002. "Phenolic Compounds and Antioxidant Capacity of Georgia-Grown Blueberries and Blackberries." *Journal of Agricultural and Food Chemistry* 50 (8): 2432–38. https://doi.org/10.1021/jf011097r.

Selma, María V., Juan C. Espín, and Francisco A. Tomás-Barberán. 2009. "Interaction between Phenolics and Gut Microbiota: Role in Human

Health." *Journal of Agricultural and Food Chemistry* 57 (15): 6485–6501. https://doi.org/10.1021/jf902107d.

Serafini, Mauro, Maria Francesca Testa, Debora Villaño, Monia Pecorari, Karin van Wieren, Elena Azzini, Ada Brambilla, and Giuseppe Maiani. 2009. "Antioxidant Activity of Blueberry Fruit Is Impaired by Association with Milk." *Free Radical Biology and Medicine* 46 (6): 769–74. https://doi.org/10.1016/j.freeradbiomed.2008.11.023.

Shukitt-Hale, Barbara, Vivian Cheng, and James A. Joseph. 2009. "Effects of Blackberries on Motor and Cognitive Function in Aged Rats." *Nutritional Neuroscience* 12 (3): 135–40. https://doi.org/10.1179/147683009X423292.

Shukitt-Hale, Barbara, Francis C. Lau, Amanda N. Carey, Rachel L. Galli, Spangler Edward L., Donald K Ingram, and James A Joseph. 2016. "*Blueberry Polyphenols Attenuate Kainic Acid-Induced Decrements in Cognition and Alter Inflammatory Gene Expression in Rat Hippocampus*" 11 (4): 172–82. https://doi.org/10.1179/147683008X301487.Blueberry.

Skrovankova, Sona, Daniela Sumczynski, Jiri Mlcek, Tunde Jurikova, and Jiri Sochor. 2015. "Bioactive Compounds and Antioxidant Activity in Different Types of Berries." *International Journal of Molecular Sciences* 16 (10): 24673–706. https://doi.org/10.3390/ijms161024673.

Smeds, Annika I., Patrik C. Eklund, and Stefan M. Willför. 2012. "Content, Composition, and Stereochemical Characterisation of Lignans in Berries and Seeds." *Food Chemistry* 134 (4): 1991–98. https://doi.org/10.1016/j.foodchem.2012.03.133.

Spencer, Jeremy P E, Katerina Vafeiadou, Robert J. Williams, and David Vauzour. 2012. "Neuroinflammation: Modulation by Flavonoids and Mechanisms of Action." *Molecular Aspects of Medicine* 33 (1): 83–97. https://doi.org/10.1016/j.mam.2011.10.016.

Spencer, Jeremy P E, David Vauzour, and Catarina Rendeiro. 2009. "Flavonoids and Cognition: The Molecular Mechanisms Underlying Their Behavioural Effects." *Archives of Biochemistry and Biophysics* 492 (1–2): 1–9. https://doi.org/10.1016/j.abb.2009.10.003.

Stalmach, Angelique, Christine A. Edwards, Jolynne D. Wightman, and Alan Crozier. 2012. "Gastrointestinal Stability and Bioavailability of (Poly)Phenolic Compounds Following Ingestion of Concord Grape Juice by Humans." *Molecular Nutrition and Food Research* 56 (3): 497–509. https://doi.org/10.1002/mnfr.201100566.

Strathearn, Katherine E., Gad G. Yousef, Mary H. Grace, Susan L. Roy, Mitali A. Tambe, Mario G. Ferruzzi, Qing Li Wu, James E. Simon, Mary Ann Lila, and Jean Christophe Rochet. 2014. "Neuroprotective Effects of Anthocyanin- and Proanthocyanidin-Rich Extracts in Cellular Models of Parkinson's Disease." *Brain Research* 1555: 60–77. https://doi.org/10.1016/j.brainres.2014.01.047.

Szajdek, Agnieszka, and E J Borowska. 2008. "Bioactive Compounds and Health-Promoting Properties of Berry Fruits: A Review." *Plant Foods and Human Nutrition* 63: 147–56. https://doi.org/10.1007/s11130-008-0097-5.

Takeuchi, Hideyuki, Shijie Jin, Jinyan Wang, Guiqin Zhang, Jun Kawanokuchi, Reiko Kuno, Yoshifumi Sonobe, Tetsuya Mizuno, and Akio Suzumura. 2006. "Tumor Necrosis Factor-α Induces Neurotoxicity via Glutamate Release from Hemichannels of Activated Microglia in an Autocrine Manner." *Journal of Biological Chemistry* 281 (30): 21362–68. https://doi.org/10.1074/jbc.M600504200.

Talavera, Séverine, Catherine Felgines, Odile Texier, Catherine Besson, Angel Gil-Izquierdo, Jean Louis Lamaison, and Christian Rémésy. 2005. "Anthocyanin Metabolism in Rats and Their Distribution to Digestive Area, Kidney, and Brain." *Journal of Agricultural and Food Chemistry* 53 (10): 3902–8. https://doi.org/10.1021/jf050145v.

Talavera, Séverine, Catherine Felgines, Odile Texier, Catherine Besson, Claudine Manach, Jean-Louis Lamaison, and Christian Rémésy. 2004. "Anthocyanins Are Efficiently Absorbed from the Small Intestine in Rats." *The Journal of Nutrition* 134 (9): 2275–79. https://doi.org/134/9/2275 [pii].

Tavares, Lucélia, Inês Figueira, Gordon J. McDougall, Helena L A Vieira, Derek Stewart, Paula M. Alves, Ricardo B. Ferreira, and Cláudia N. Santos. 2013. "Neuroprotective Effects of Digested Polyphenols from

Wild Blackberry Species." *European Journal of Nutrition* 52 (1): 225–36. https://doi.org/10.1007/s00394-012-0307-7.

Tungmunnithum, Duangjai, Areeya Thongboonyou, Apinan Pholboon, and Aujana Yangsabai. 2018. "Flavonoids and Other Phenolic Compounds from Medicinal Plants for Pharmaceutical and Medical Aspects: An Overview." *Medicines* 5 (3): 93. https://doi.org/10.3390/medicines5030093.

Van De Velde Franco, María E Pirovani, and Silvina R Drago. 2018. "PT." *Journal of Food Composition and Analysis*. https://doi.org/10.1016/j.jfca.2018.05.007.

Vitaglione, Paola, Giovanna Donnarumma, Aurora Napolitano, Fabio Galvano, Assunta Gallo, Luca Scalfi, and Vincenzo Fogliano. 2007. "Protocatechuic Acid Is the Major Human Metabolite of Cyanidin-Glucosides." *The Journal of Nutrition* 137 (9): 2043–48. https://doi.org/10.1093/jn/137.9.2043.

Walsh, Dominic M., Igor Klyubin, Julia V. Fadeeva, William K. Cullen, Roger Anwyl, Michael S. Wolfe, Michael J. Rowan, and Dennis J. Selkoe. 2002. "Naturally Secreted Oligomers of Amyloid β Protein Potently Inhibit Hippocampal Long-Term Potentiation in Vivo." *Nature* 416 (6880): 535–39. https://doi.org/10.1038/416535a.

Walton, Michaela C., Tony K. McGhie, Gordon W. Reynolds, and Wouter H. Hendriks. 2006. "The Flavonol Quercetin-3-Glucoside Inhibits Cyanidin-3-Glucoside Absorption in Vitro." *Journal of Agricultural and Food Chemistry* 54 (13): 4913–20. https://doi.org/10.1021/jf0607922.

Wang, Dongjie, Lap Ho, Jeremiah Faith, Kenjiro Ono, Elsa M Janle, Pamela J Lachcik, Bruce R Cooper, et al., 2015. "*Role of Intestinal Microbiota in the Generation of Polyphenol-Derived Phenolic Acid Mediated Attenuation of Alzheimer's Disease -Amyloid Oligomerization*," 1025–40. https://doi.org/10.1002/mnfr.201400544.

Wang, Jun, Weina Bi, Alice Cheng, Daniel Freire, Prashant Vempati, Wei Zhao, Bing Gong, et al., 2014. "Targeting Multiple Pathogenic Mechanisms with Polyphenols for the Treatment of Alzheimer's Disease-Experimental Approach and Therapeutic Implications."

Frontiers in Aging Neuroscience 6 (MAR): 1–10. https://doi.org/10.3389/fnagi.2014.00042.

Williams, Claire M., Manal Abd El Mohsen, David Vauzour, Catarina Rendeiro, Laurie T. Butler, Judi A. Ellis, Matthew Whiteman, and Jeremy P.E. Spencer. 2008. "Blueberry-Induced Changes in Spatial Working Memory Correlate with Changes in Hippocampal CREB Phosphorylation and Brain-Derived Neurotrophic Factor (BDNF) Levels." *Free Radical Biology and Medicine* 45 (3): 295–305. https://doi.org/10.1016/j.freeradbiomed.2008.04.008.

Williams, Robert J., and Jeremy P E Spencer. 2012. "Flavonoids, Cognition, and Dementia: Actions, Mechanisms, and Potential Therapeutic Utility for Alzheimer Disease." *Free Radical Biology and Medicine* 52 (1): 35–45. https://doi.org/10.1016/j.freeradbiomed.2011.09.010.

Youdim, Kuresh A., Michael S. Dobbie, Gunter Kuhnle, Anna R. Proteggente, N. Joan Abbott, and Catherine Rice-Evans. 2003. "Interaction between Flavonoids and the Blood-Brain Barrier: In Vitro Studies." *Journal of Neurochemistry* 85 (1): 180–92. https://doi.org/10.1046/j.1471-4159.2003.01652.x.

In: Rubus: An Overview
Editor: Davet Chabot

ISBN: 978-1-53617-376-5

Chapter 3

PHYTOPATHOLOGY GENERALITIES IN THE GENUS *RUBUS*: A REVIEW

D. Verdugo[1], A. Vargas[1], D. Gajardo[1], V. D'Afonseca[2], Ph D, S. Espinoza[3], Ph D and G. González[4,*], Ph D

[1]Ingeniería en Biotecnología, Facultad de Ciencias Agrarias y Forestales, Universidad Católica del Maule, Talca, Chile
[2]Vicerrectoría de Investigación y Posgrado, Universidad Católica del Maule, Talca, Chile
[3]Departamento de Ciencias Forestales, Facultad de Ciencias Agrarias y Forestales, Universidad Católica del Maule, Talca, Chile
[4]Centro de Biotecnología de los Recursos Naturales, Facultad de Ciencias Agrarias y Forestales, Universidad Católica del Maule, Talca, Chile

ABSTRACT

In marketable crops, fruit characteristics such as taste, size, and color must be adjusted to consumer demands. This positions the species from the

* Corresponding Author's Email: ggonzalez@ucm.cl.

genus *Rubus* with interesting perspectives due to its versatility to satisfy different markets. Species of this genus have been selected and backcrossed to obtain homogeneous crops with fruit that fits the demanded characteristics; however, after repeated cycles of selection and breeding, the species's genetic variability decreases and its susceptibility to pathogens increases, diminishing the crops' viability, vigor, and yield.

The genus *Rubus* includes several varieties differing in maturity, productivity, and with a variable tolerance to different diseases. Because of this, growers are constantly seeking for new varieties with better organoleptic qualities and high resistance to diseases. Unfortunately, a number of diseases associated to more than 740 species of bacteria, fungi, virus, and nematode agents have been described throughout the world, mainly affecting fruit quality in varieties of the genus *Rubus*. Those pathogens cause tissue necrosis, atrophy or deformation of organs, wilt, fruit crumbling and dwarfing, galls, root damage, and postharvest damage among other diseases.

This book chapter is aimed to review the main pathogens and phytoparasites that cause damage in raspberry (*Rubus idaeus*) causing important economic losses in the crop. We also present a synthesiezed view of the various etiological agents that affect the genus *Rubus*.

Keywords: raspberry, diseases, virus, fungi, bacteria, nematodes

INTRODUCTION

The genus *Rubus*, commonly known as berries, harbors thousands of native species distributed wolrwide (Sochor, et al., 2018). Species of this genus, such as raspberries (*Rubus idaeus*) and blackberries (*Rubus alleghensis*, *Rubus ursinus*) are commercialized in the global marketplace due to their high profitability and value in health. Currently, those species have captured the attention of the nutraceutical industry due to their high content of a variety of bioactive compounds such as antioxidants, anti-inflammatories, chemo-preventives and antimicrobial molecules that corroborates their potential uses for human health (Schulz and Chim, 2019).

In 2005, the global blackberry production was estimated to be around 140,292 tons (Strik et al., 2008). Due to the high demand for raspberry fruit, the agricultural industry aimed to increase the production yields and improve the crops' uniformity, focusing in fruit characteristics such as color, flavor, size, nutrient content, and durability. Species with desired qualities by the consumer are selected and breed in order to obtain new varieties that produce quality fruits. However, one point of vulnerability in this type of approach is the reduction of genetic variability, which has increased the frequency of genotypes susceptible to pathogens. This has caused an increase in losses for raspberry and blackberry crops that have reached USD 39.4 million from 2009 to 2014 (Farnsworth et al., 2017; Graham and Brennan, 2018).

This last point represents one of the main concerns for producers, forcing them to look for new techniques or products capable of improving resistance to pests and diseases caused by virus, fungi, bacteria, and nematodes, including pesticides, biocontrollers, and genetically modified organisms. Thus, understanding the mechanisms of action of those pathogens will provide relevant information to its control and eradication. In this book chapter, we synthesize information about the main bacteria, fungi, virus, and nematode agents that affect the genus *Rubus*.

FUNGI IN THE GENUS *RUBUS*

In agriculture, phytopathogen fungi are the primary cause of pre and post- harvest diseases that threaten production and people's health, and are responsible for important economic losses. Fungi induce biological changes and cause negative effects on growth and development of the host species, being the primary cause of the pathologic deterioration of fruits, leaves, stems, and roots.

The genus *Rubus* have a great export potential and economic importance; however, crops are exposed to fungi diseases and fruits have a limited post-harvest storage time due to an increase of their metabolism after harvesting, experiencing organoleptic changes and appearance modifications that are followed by an increase in susceptibility to rot fungi (Candan, 2006).

Peronospora sp.

Important damages such as fruit dehydration, severe malformations, lack of brightness, and the detention of the maturation process have been reported for *Rubus glaucus*, associated to *Peonospora* sp. (Hincapié et al., 2017), causing loss of production in the 50 to 70% range (Rodríguez et al., 2017). The use of fungicides contributes to the development of more resistant strains, which are difficult to control (Hukkanen et al., 2006). In countries such as Colombia *Peronospora* incidence is a recurrent problem in *R. glaucus* causing controversy in its identification. Tamayo (2001), Hincapié et al. (2017) and the Manual del Cultivo de Mora de la Gobernación de Antioquia (2014), pointed out that the causing agent of the powdery mildew is *Peronospora corda*. Hurtado (2019) suggests that *Peronospora rubi* is the causing agent, but ICA (2011) and Montoya et al. (2003) proposes that *Peronospora sparsa* is the responsible organism.

Peronospora corda also causes the sooty mold disease which affect the stems, flowers and fruits. The affected stems show purple discolorations with whitish injuries on which a light gray color begins to grow (Saldarriaga and Bernal, 2000; Tamayo, 2003). The flowers show a yellowish hue and then their petals dry up. On the fruit, the fungus causes an irregular development, uneven maturation, and a lack of brightness, which provokes losses in its commercialization that goes from to 20 to 30% of the harvested fruit (Tamayo and Peláez, 2000).

Fusarium oxysporum

Fusarium oxysporum is an asexual fungus distributed worldwide that can be spread by air, seeds, and infected material (Garibaldi et al., 2004). It has a high number of hosts and causes great economic losses, being considered as the most important species of the genus Fusarium (Leslie and Summerell, 2006). *Fusarium oxysporum* f. sp. *fragariae* causes fusarium wilt on strawberry plants. It was first described in Australia (Winks and Williams, 1965), but it has also been detected in Argentina (Mena et al.,

1975; Baino et al. 2010), Korea (Cho and Moon, 1984; Nam et al, 2011), Chile (González et al. 2005), China (Zhao et al., 2009) and the United States (Williamson et al., 2012). The pathogen resides on the soil and behave as either saprophytic, decomposing the lignin and the carbohydrate complexes associated to organic soil rests (Rodríguez et al., 1996; Christakopoulos et al. 1996); or endophytic, colonizing the plant roots (Katan, 1971; Gordon and Martyn, 1997). This fungus can also protect the plant or be the base of suppression for other diseases (Lemanceau et al., 1993; Oyarzun et al., 1994).

The effect of the biotic factors on the genus *Rubus* has been evaluated in the bioactive compound content (Ramos et al., 2014; 2015). For *Rubus fructosus* it has been found that the isolated disaccharide of *F. oxyporum* promotes the activity of the phenylalanine ammonia lyase enzyme in its cells at nanomolar concentrations (Nita Lazar et al., 2004). This suggests that *R. glaucus* could be responding on a similar way to the fungi infection.

The morphologic characteristics of the *F. oxysporum* species includes the production of microconidia in false heads over short phialides formed from hyphae, the Chlamydospores production (Resistence of structure), and the presence of Macroconidia (Leslie and Summerell, 2006). In strawberry cultivation, the symptoms of wilt produced by Fusarium are fostered by moderately high temperatures (>22ºC). In plants inoculated with *F. oxysporum*, an increase in temperature from 17 to 27°C significantly increases disease severity (Fang et al. 2011). Wilt development due to exposure to *F. oxysporum* has also been associated to factors such as soil water deficit, presence of plagues and nematodes, and plant stress in the moments of maximum production. The disease caused by *F. oxysporum* is very hard to control due to the long survival of the chlamydospores in the soil, its wide host range and the ineffectiveness of the existing fungicides. For this reason, there is currently no effective treatment to control this pathogen and most of the efforts are oriented to prevent the disease.

In many crops there are genes capable of providing an effective resistance to pathogens, but deployment of new varieties with pathogen resistance is challenged by new fungi strains that overcome this resistance (Jiménez-Gasco et al., 2004). Thus, biological control is a good alternative for Fusarium diseases. This is the case of Bacillus- and Trichoderma-based biofungicides that have been proved to reduce fusarium wilt symptoms (Pastrana, 2014). Non-pathogenic strains of *F. oxysporum* or others antagonist microorganisms have also been referred as biocontrol agents (Fravel et al., 2003). Since all of them derive from nature, they have less environmental impact than the traditional fungicides.

Macrophomina phaseolina

Macrophomina phaseolina is a soil fungus distributed worldwide causing crop losses in Turkey (Benlioğlu et al., 2004), Israel (Zveibil and Freeman, 2005), United States (Mertely et al., 2005; Koike, 2008), Australia (Golzar et al., 2007; Fang et al., 2011), Spain (Avilés et al., 2008), Iran (Sharifi and Mahdavi, 2011), Argentina (Baino et al., 2011) and Chile (Sánchez et al., 2013).

This pathogen is the main cause of the charcoal rot disease and it is present in two states: 1) picnidic (*M. phaseolina*), and 2) sclerocial (*Sclerotium bataticola*), which is more frequent. The picnidic state has been observed in diverse tissues of its host plants. However, it is not too frequent and its identification is usually confounded with the sclerocial state (Bascón, 2009). In the sclerocial state, the sclerotia survive in the soil and on residual biomass obtained at the end of the growing season, and remain on the soil forming a primary inoculum for the new crop (Zveibel et al., 2012). Some authors have described a reduction in the number of sclerotia on soils that have been exposed to wet periods (Pratt, 2006), but some others have concluded they are capable of resisting in the soil up to three years in sufficient quantities to ensure the infection of subsequent crops (Dhingra and Sinclair, 1975). The sclerotia germination is produced on the root surface when temperatures vary between 28 and 35°C. The germ tubes form an apressoria capable of penetrating the walls of the epidermic cells through mechanic pressure and enzymatic digestion or through natural openings (Bowers and Russin, 1999). Once inside the plant, the hyphae start to grow intercellularly through the cortex and then intercellularly through the xylem, colonizing the vascular tissues. This way, *M. phaseolina* spreads through the root and the base of the stem causing microsclerotia that clog vessels (Wyllie, 1988).

Phytophthora cactorum, *Phytophthora fragariae*

The first symptoms of these two pathogens are necrosis on the edge of the leaves, followed by wilt and the death of leaf apex, along with chlorotic and wilt of lateral outbreaks. The necrosis progresses until the plants' shoot dies. By examining the plants' roots and root collar, it is appreciated necrosis and the detachment of the fundamental epidermis, under which a brownish and reddish coloration is produced. The affected plants produces less buds with less vigor and symptoms of lack of nutrients, caused by the damaged roots. While the disease progresses in the orchard, the plant's population diminishes and weeds cover the orchard gradually. The affected plants are more susceptible to frost and become less productive; their fruits have more acidity and usually die prematurely (Undurraga and Vargas, 2013).

Pseudocercospora and *Ramularia*

The *Pseudocercospora* species are well known as pathogens in a wide variety of ornamental plants and food crops, predominantly in tropical and subtropical environments where they cause leafy spots, blights, stains, and fruit rot. Some of these species are used in biological weed control (Den Breeÿen et al., 2006). It has been reported that some species are found on different hosts that belong to a single plant family (Deighton, 1976), which is contrasted to an initial hypothesis that most of the *Pseudocercosporas* were strictly specific to one host.

Until recently, *Pseudocercospora* was treated as an anamorphic genus associated to the *Mycosphaerella* species (*Mycosphaerellaceae, Capnodiales*), along with approximately other thirty anamorphic genera

(Crous, 2009). The name *Mycosphaerella* is restricted to species with anamorphs of *Ramularia*, being this a name older than *Mycosphaerella*. Nowadays, only a generic name is used to name the species of *Pseudocercospora* (Hawksworth et al., 2011; Wingfield et al., 2011).

Verticillium dahliae, Verticillium albo-atrum

Verticillium wilt is caused by *Verticillium* fungi and it is present in all the raspberry productive zones, although it causes more losses in irrigated arid and semi-arid areas. There are two types of *Verticillium* pathogens in raspberries crops namely *Verticillium dahliae* and *Verticillium albo-atrum.* These species have a wide spectrum of plant hosts including raspberry, tomato, cotton, perennial crops, and diverse weeds (Mass, 1998). The affected plants shows chlorosis and foliage wilt in summer; even in well watered plants, due to clogging of the xylem. Wilt disappears during the night or during cloudy days, but it appears again with high temperature, until the leaves and buds dry up. The production decreases and many fruits do not reach ripeness or are more acid. These symptoms can be confused with those of the *Phytophthora* species.

Botrytis cinerea

This pathogen causes the gray mold rot disease, which is one of the most common and serious diseases that affects the different productive zones of the genus *Rubus* in countries such as the United States and China (McNicol et al., 1985; Castro et al., 1995; Hincapié, 2017). It affects strawberries (Jarvis, 1977; Coley-Smith et al., 1980), grapes (Nelson, 1956), and raspberries (Williamson et al., 1987) at the worldwide level. It is presented as white cottony mycelia which changes to a brownish color when it turns into a sporula. In the fruit, at the beginning of the infection, it manifests as a watery putrefaction affecting only part of the achene that decolorates and sinks. After that, the white mycelium emerges and can acquire its typical ashen coloration (Jennings, 1988; Pearson and Goheen, 1996). The most important fruit infections take place in greenhouses because of the high humidity. This suggests that greenhouse ventilation systems could effectively reduce the severity of gray mold rot in raspberry fruits.

Gray mold rot also appears during the production and post harvest phases. This fungus affects the flower buds since its blossom and appears in the fructification and maturation, causing necrosis and fruit mummification (Hincapié, 2017; Quinatoa, 2015). In the presence of humidity, the fungus grows with a variety of colors including gray, brown or olive green. The fruits dry up, get mummified and adhered to the bunch (Saldarriaga and Bernal, 2000; Tamayo, 2009). This disease has produced losses between 50 and 76% of the harvested fruit (Tamayo and Peláez, 2000). The fungus can also affect the leaves, flowers and stems (Franco and Giraldo, 2000).

Other diseases that affect the genus *Rubus* are the oidium and the yellow rust. The oidium disease is caused by *Sphaerotheca macularis* and the infection appears mainly in young leaves through deformation (i.e., leaves curvature). It is associated to the presence of chlorotic, irregular, and diffused areas that can be seen on the leaf surface. Occasionally, the leaves are covered with a fine white dust that corresponds to the growth and the sporulation of the fungus. The yellow rust is caused by *Gerwagsia lagerheimii*. It appears in the form of pustules with orange color on the stems, the lower leaf surface and the fruits. There are also reports of foliage infections caused by *Septoria* sp, *Phyllosticta* sp, and *Alternaria* sp, whose importance is still secondary due to the low incidence on production. The stems and roots of *Rubus ulmifolius* also present some diseases caused by fungus like *Coniothyrium fuckelii*, *Verticillium albo-atrum*, *Rosellinia* sp, *Fusarium roseum* and *Fusarium oxysporum*, whose importance has not been established yet (Tamayo, 2009).

PROKARYOTES IN THE GENUS *RUBUS*

All prokaryotes are single-celled microorganisms, lacking internal compartments, having a cell membrane, cytoplasmic ribosome 70S and a nuclear region not delimited by membrane (Trigiano et al., 2007). Prokaryotes can be taxonomically classified based on distinctive characteristics in their morphology (i.e., presence or absence of cell wall), photopigments, biochemistry, tolerance to extreme living conditions and molecular analyzes (Kämpfer and Glaeser 2013). Pathogenic bacteria can produce devastating effects on plant productivity and yield, being a threat to agricultural practices and affecting various economic areas (Martins et al. 2018). The soil is the environment of thousands of these microorganisms that can interact with eukaryotic organisms, exhibiting strong adaptability, a versatile metabolism and ingenious mechanisms designed to modify the development of their hosts; in the particular case, the plants, making them a source of nutrition and refuge.

The prokaryotes agents' causative of plant diseases have the ability to colonize and to multiply in vegetal tissues by increasing their population size from a small cell amount to high number of cells in a remarkably short period of time (Trigiano et al., 2007). The entry of microrganisms into the host plant can be through openings in the natural physical barriers such as stomata, hydathodes, and lenticels, or by wounds originated during root growth or leaf abcision. When this occurs, signals are produced to indicate that the pathogen has entered trhough passive invasion. When the natural barriers of the plants are damaged, due the action of enzymes such as cutinases and esterases (Ban et al., 2008); or alterations in hydrostatic pressure, it is called active invasion (Trigiano et al, 2007).

An important group of prokaryotes are the mollicutes, bacteria that lack a cell wall and are usually intracellular parasites of animal and plant organisms (Bertaccini and Duduk 2009; Trigiano et al, 2007). The mollicutes (Table 1), called phytoplasms, will be described along with other pathogens that affect the species of genus *Rubus*.

Table 2. Prokaryotes that infect plants of genus *Rubus*

Species	Host	Disease	Reference
Erwinia amylovora	Rubus sp	Fireblight	(Mann et al. 2012)
Agrobacterium tumefaciens	*R. idaeus*	Crown gall	(Undurraga and Vargas 2013)
Agrobacterium arsenijevicii	*R. idaeus*	Crown gall	(Kuzmanović et al. 2015)
Agrobacterium rosae	*R. idaeus*	Tumors	(Kuzmanović et al. 2018)
Agrobacterium rubii	*Rubus* spp	Crown gall	(Undurraga and Vargas 2013)
Candidatus Phytoplasma rubi	*Rubus* spp	Rubus stunt	(Davies, 2000; Linck, 2019)

Erwinia amylovora

Erwinia amylovora is a gram-negative phytopathogen whose distribution includes North America, New Zealand, Europe and the Middle East. It can be classified into two groups. The first group harbor strains with small genetic diversity that are capable of infecting various members of the *Spiraeoideae* subfamily such as apple, pear, cotoneaster, hawthorn, and quince. The second group presents a greater genetic diversity with strains that are capable of infecting the genus *Rubus*, causing the fireblight disease. The dispersion of *E. amylovora* occurs by insects, splashes of rain on a local scale, host plants with a latent infection or with undetectable canker sores (Mann et al., 2012). Generally, this pathogen causes wilting of leaves, green or brown shoots, or black shoots in organs with an advanced infection. Additionally, exudates and canker sores eventually leads to the plant's death (Bühlmann et al., 2013, Bartho et al., 2019).

This bacterium enter into a plant through nectaries in flowers or wounds caused by hail, storms, or insects, and it can spread rapidly into the plant vascular system. One notable way for its spread is the Type III secretion system, which can secrete various proteins that inactivate the host's defenses. Among them, AvrRpt2EA, an effector protein that acts as a cysteine protease activated by cyclophilin, is a key molecule in infection processes in plants (Bartho et al., 2019). Other factors that help for a successful infection of *E. amylovora* are proteins involved in biosynthesis of exopolysaccharides and siderophores-mediated iron uptake systems (Mann et al., 2012).

Agrobacterium tumefaciens

The genus *Agrobacterium* is characterized by a specific coregenome architecture consisting of a circular chromosome and a large linear chromate. This genus holds groups of bacteria that reside in soils and plants; known as phytopathogens, and causes neoplastic diseases in raspberry and various plants and agricultural crops (Kuzmanović et al. 2018).

A recognized representative among the phytopathogens of the genus *Rubus* is *Agrobacterium tumefaciens*, a gram-negative bacterium and etiology agent of crown gall disease that lives in soil and water. Injuries in the plant tissue are provided through which *A. tumefaciens* can enter the plant. Chemotaxis induced by the production of asytosyringone during tissue injury helps this bacterium enters into the plant through wounds in the root system, coming in contact with the cells of the root endodermis and introducing part of its genetic material. This bacterium has the ability to insert its genetic material into the host plant genome through the named *Tumor Inducing Plasmid* (Ti plasmid) (Guo et al., 2019). The Ti plasmid carriers about 25 virulence genes distributed in seven operons that are integrated into the host genome and, once transferred, they can express in infected cells and induce tumors. The plasmid also harbors genes encoding plant hormone synthesis enzymes (i.e., cytokinin and auxin) and opine. The accumulation of these phytohormones in the transformed host cells causes cellular hyperproliferation that leads to the neoplastic growth of the transformed tissue. In addition, opine production creates a specific ecological niche for *A. tumefaciens* (Guo et al., 2011).

Candidatus Phytoplasma rubi

Phytoplasms are obligatory phytopathogenic bacteria belonging to mollicutes that colonize the phloem of their host plants. These bacteria are transferred into the plant by the sap-feeding insect *Macropsis fuscula* or by vegetative propagation. In the phloem, the phytoplasms are reporduced in the sieve tubes by binary fission, budding or fragmentation and then it spreads to the whole plant. The lack of metabolic pathways to synthesize its own nutrients make *C. Phytoplasma* rubi to obtain them from the host, resulting in unbalanced host biomolecular levels, and generating yellowing, chlorosis, wilting, dwarfism, and bud proliferation in the infected plant (Weintraub and Jones, 2010). In *Rubus* species they cause a disease called Rubus stunt (Davies, 2000). Symptoms of Rubus stunt include proliferation of axillary buds, stunted growth, witch's broom disease, small leaves, short internodes, enlarged sepals, phyloids, flower proliferation, and fruit malformations (Davies, 2000; Mäurer and Seemüller, 1995). However, although these symptoms are similar in all species, in some cases no characteristic signs of proliferation are observed (Davies, 2000) because the symptomatology is strongly influenced by environmental factors, agronomic characteristics, disease progression or mixed infections with viruses (Ermacora and Osler, 2019). All these factors make the visual detection of the disease a difficult task (Linck et al., 2019). Rubus stunt cannot be detected early because the symptoms of this disease are not visible during the latency period or because of the above-mentioned factors. Therefore, it is necessary its detection by molecular tools such as Nested PCR or Real-time multiplex PCR assays before conducting vegetative propagation in order to avoid yield losses in various economic sectors (Linck et al., 2016). Formerly, it was thought that the diseases associated with these pathogens were caused by viruses because phytoplasms are not cultivable in artificial media and they are difficult to obtain in pure media and have incubation periods of up to one year (Reveles-Torres et al., 2014).

VIRUS IN THE GENUS *RUBUS*

Some plant genera, such as *Rubus* are hosts for a number of viruses that produces important economic losses (Smith, 1948; Quito-Avila et al., 2014). Infected *Rubus* plants presents symptoms such as appearance of rings, veins clearing, chlorosis, stunted growth, fruit crumbling, decrease the number of fruits, distortion of leaf margins, deformed fruit, rough young leaves, annular leaf spots, mosaic, mottled appeareance, necrosis, leaf line patterns, reduced plant vigor, early senescence, reduced size of the fruit, and twisting of leaves.

Viruses are efficient pathogen agents of small size that can cause high decay in an infected plant. They are made of a tiny fraction of nucleic acid wrapped into a protein coat (or capsid) that protects the viral genome, and are inactive outside the host cells. Because of this, in many cases the virus are not considered living organisms (Gergerich, 2007). Some viruses are covered by an outer membrane composed of lipids and proteins, while the capsids are assembled following two fundamental types of symmetry. The first type is the icosahedral structure that has two basic forms: the bacilliform virions and the twin virions. In both, the genomic nucleic acid forms a partially organized sphere inside the protein capsid. The second type is helical, which has two variants, rigid rods and flexible filaments. In both, the nucleic acid is highly organized. They take the same helical conformation like the capsid protein.

Table 3. Species of virus that infect plants of Rubus genus

Virus name	Acronym	Mode of transmission	Genus
Beet pseudo yellows	BPYV	Greenhouse whitefly	*Crinivirus*
Blackberry chlorotic ringspot	BCRV	Pollen, seed	*Ilavirus*
Blackberry virus S	BlVS	-	*Marafivirus*
Blackberry virus Y	BVY	-	*Brambyvirus*
Cherry rasp leaf	CRLV	Nematode	*Cheravirus*
Impatiens necrotic spot	INSV	Thrips	*Tospovirus*
Raspberry bushy dwarf	RBDV	Pollen, seed	*Idaeovirus*
Raspberry leaf blotch	RLBV	Mites	*Emaravirus*
Raspberry leaf curl	RpLCV	Aphid	No info
Raspberry leaf mottle	RLMV	Aphid	*Closterovirus*
Raspberry ringspot	RpRSV	Nematode, pollen, seed	*Nepovirus*
Raspberry vein chlorosis	RVCV	Aphid	*Rhabdovirus*
Rubus canadensis virus 1	RuCV-1	-	*Foveavirus*
Rubus yellow net	RYNV	Aphid	*Badnavirus*
Sowbane mosaic virus	SoMV	Pollen, seed	*Sobemovirus*
Tomato black ring	TBRV	Nematode, pollen, seed	*Nepovirus*
Tomato ringspot virus	ToRSV	Nematode, pollen, seed	*Nepovirus*

Viruses have various propagation methods, but transmission through vectors is one of the most common. Vectors generally can be insects of the Amphorophora, Aphididae and Pseudococcidae families; which, by feeding on different plants, transfer the infection from an infected plant to a non-infected plant (Diaz-Lara et al., 2015; Jones and Roberts, 1976; Tidona and Darai, 2011). Another common vector are the nematodes, which are present in the soil and can transmit the virus through the roots, with the plant showing symptoms in spots of variable size, reflecting the distribution of the nematode (Murant et al., 1996). The most prominent viruses that have been reported infecting raspberry plants are summarized in Table 2.

Rubus Yellow Net Virus (RYNV)

RYNV is a circular double-stranded DNA virus (size ~ 7,500 bp) belonging to the genus *Badnavirus* which presents a bacilliform-shaped virus particle with dimensions of 80-150 × 25-30 nm (Kalischuk et al. 2013; Swiss Institute of Bioinformatics, 2019). RYNV-infected plants present structural alterations, especially in limbo cells where it produces a rare growth of the cell wall (Jones and Roberts, 1976). This change leads to chlorosis that occurs along the venous network (Raspberry Vein Mosaic disease).

Blackberry Virus y (BVY)

BVY is a positive-sense linear ssRNA virus (size ~ 11 kb) belonging to the genus *Brambyvirus*. This virus has an 800 × 11–15 nm flexuous filament viral particle (Swiss Institute of Bioinformatics 2019). This is the only specimen of its genus and it was identified in blackberry. It is also able to infect raspberry cv. Meeker asymptomatically, event seen in areas of the United States, and it is also presumed to act in the presence of other viruses (Susaimuthu et al., 2008).

Cherry Rasp Leaf Virus (CRLV)

CRLV is a positive-sense linear ssRNA virus of the genus *Cheravirus* with an icosahedral viral particle of size about 25-30 nm (Swiss Institute of Bioinformatics, 2019). CRLV-infected plants may have stunted growth (James et al., 2001).

Raspberry Leaf Mottle (RLMV)

RLMV is a positive-sense linear ssRNA virus belonging to the genus *Closterovirus* that presents a non-enveloped, flexible and unusually long filamentous viral particle of approximately 1250-2200 nm in length and 10-13 nm in diameter (Swiss Institute of Bioinformatics, 2019). This virus is responsible for the raspberry mosaic disease and fruit crumbling (Quito-Avila and Martin, 2012).

Beet Pseudo Yellows (BPYV)

BPYV is a positive-sense linear ssRNA virus, member of the genus *Crinivirus* that present a non-enveloped bipartite filamentous viral particle of approximately 650-850 nm and 700-900 nm in length and 10-13 nm in diameter (Swiss Institute of Bioinformatics, 2019). The presence of this virus is associated with symptoms that include severe yellowing, reduced fruit size and possibly early senescence. The symptoms begin with the manifestation of chlorotic intervenous areas in the leaves, which spread until the majority of the leaves are chlorotic and the veins remain green. Symptomatic leaves are more fragile and with a thick appearance (Wintermantel, 2004). BPYV is transmitted by the semiperistant glasshouse whitefly (*Trialeurodes vaporariorum*) (Boubourakas et al., 2006).

Raspberry Leaf Blotch (RLBV)

RLBV is a segmented RNA virus that presents its nucleic acid into four sections of linear, negative and single-stranded RNA. This virus is member of the genus *Emaravirus* and presents a spherically encapsulated viral information in a complete genome of approximately 12.2 kb of size (Swiss Institute of Bioinformatics, 2019). This pathogen is associated with the Raspberry Leaf Spot Disease wich presents symptoms such as large yellow spots or rings in the leaves, leaves malformation, severe necrosis, and reduction of plant vigor (Dong et al., 2016).

Rubus Canadensis Virus 1 (RuCV-1)

RuCV-1 is a positive-sense linear ssRNA virus belonging to the genus *Foveavirus* with a non-enveloped, flexible, filamentous viral particle of about 800 nm of longitude and 12-13 nm in diameter, and a genome size of 8.4-9.3 kb (Swiss Institute of Bioinformatics, 2019). The virus is associated with symptoms of mild chlorosis and vein clearing (Abou Ghanem-Sabanadzovic et al., 2013).

Raspberry Bushy Dwarf Virus (RBDV)

RBDV is a positive tripartite segmented linear ssRNA virus, member of the genus *Idaeovirus* that presents an isometric non-enveloped viral particle of 33 nm of size. It has three RNA components (2 genomic and 1 subgenomic (sgRNA)), with sizes of RNA1= 5.4 kb, RNA2= 2.2 kb, and sgRNA3= 1 kb (Swiss Institute of Bioinformatics, 2019). The main symptom of this virus is the crumbly fruit that generates great losses to the farmers. Another main characteristic of this virus is that it is transmitted through infected pollen, which makes its containment very difficult once this virus is present in mature infected plants (Ward et al., 2012).

Blackberry Chlorotic Ringspot Virus (BCRV)

BCRV is a positive tripartite linear ssRNA virus, belonging to the genus *Ilarvirus*. This virus has a non-enveloped spherical viral particle of approximately 29 nm in diameter with an icosahedral symmetry. Additionally, it has four RNA components, 3 genomic and 1 subgenomic (sgRNA), with sizes of RNA1= 3.4 kb, RNA2= 3.1 kb, RNA3= 2.2 kb, and sgRNA= 1 kb (Swiss Institute of Bioinformatics 2019). The main symptoms of this virus are diffuse chlorotic spotting and ringspots in leaves (Jones et al., 2006).

Blackberry Virus S (BIVS)

BIVS is a positive-sense linear ssRNA virus of the genus *Marafivirus* with an isometric non-wrapped capsid viral particle and with icosahedral symmetry of approximately 30 nm in diameter. It presents a genome size of 6-7 kb (Swiss Institute of Bioinformatics, 2019). The BIVS-infected plants show mosaic and mottled yellow vein symptoms (Sabanadzovic and Abou Ghanem-Sabanadzovic, 2009).

Tomato Ringspot Virus (ToRSV)

ToRSV is a positive-sense bipartite linear ssRNA virus member of the genus *Nepovirus*. It presents two viral particles. Particle B contains 7.5 kb RNA1 in size and particle M contains RNA2 3.9 kb in size (Swiss Institute of Bioinformatics, 2019). Symptoms of ToRSV are marked rings and chlorotic designs accompanied by vein whitening that usually occurs during spring and tends to disappear during the summer. In raspberry plants the viral infection is often asymptomatic; however, plants present low vigor or yield and/or fruit deformation (González Silva et al., 2017).

Raspberry Vein Chlorosis Virus (RVCV)

RVCV is a linear negative-sense ssRNA virus, member of the family *Rhabdoviridae*. Its viral particle is enveloped with a bacilliform shape of 180 nm long and 75 nm of wide. This virus present a viral RNA about 11-15 kb in size (Swiss Institute of Bioinformatics, 2019), and is the causative agent of raspberry venous chlorosis disease. The visible symptoms of the disease are a minor vein chlorosis; which appears as a fine network, with possible distortion of the leaf lamina. In some cases, the virus causes a significant reduction in fruit yield, and can cause thinning of the stems and failure at the beginning of the fruit formation (Mcgavin and Macfarlane, 2011).

Sowbane Mosaic Virus (SoMV)

SoMV is a linear, monopartite, positive-sense ssRNA virus, belonging to the genus *Sobemovirus*. This pathogen presents an uncoated viral particle with icosahedral symmetry of 30 nm in diameter. Its viral RNA is about 4 kb in size (Swiss Institute of Bioinformatics, 2019). Symptoms of SoMV in infected plants are pale chlorotic spots and rough young leaves with pronounced sinus in the leaf margins (Eastwell et al., 2010).

Impatiens Necrotic Spot Virus (INSV)

INSV is a linear negative sense ssRNA virus segmented into three parts (L, M, S) and it is a member of the genus *Tospovirus*. This virus presents a spherical enveloped viral particle of 80-120 nm in diameter. Its nucleic acids are segmented and presents different sizes. The L, M, and S segments are approximately 8.8, 4.8 and 3 kb in size (Swiss Institute of Bioinformatics, 2019). Symptoms of INSV in infected plants are a chlorotic mottled leaf accompanied by leaf necrosis (Kondo et al., 2011).

Raspberry Ringspot Virus (RpRSV)

RpRSV belongs to the genus *Nepovirus* and shares the same characteristics of the members of its genus. The symptoms of infected plants are the appearance of chlorotic rings in the leaves (Diaz-Lara et al., 2019).

Tomato Black Ring Virus (TBRV)

TBRV also belongs to the genus *Nepovirus* and shares the same characteristics of the members of its genus. Symptoms of TBRV in infected plants are systemic chlorotic ring spot, yellowing and delayed vein growth, mottling and deformation of the leaves and necrotic spots (Harper et al., 2011; Brunt et al., 1996).

Raspberry Leaf Curl Virus (RpLCV)

RpLCV is a virus that has not yet a genetic characterization and its symptoms only appear in the following season after the infection where leaves of the primocanes and the floricans are severely curly, distorted and chlorotic. Additionally, the buds are very small and can branch out greatly in the nodes. The fruits of infected plants are small and crumbly. Severly infected plants dramatically reduces its yield, ceasing to have an economic value. However, it does not influence the total production because of the low spread of the disease (Martin et al., 2013).

In nature, it is not possible to find only species infected with a single type of virus. The co-occurrence of at least three viruses in raspberries has been found. This is an important issue to know since plants can show several symptoms or can camouflage the presence of other viruses, which compromises the control plans (Medina et al., 2006).

NEMATODES IN THE GENUS *RUBUS*

Nematodes are multicelled organisms that inhabit aquatic and terrestrial environments. They are cylindrical worms with free life and parasitic life habits. The latter form of life affects animals and plants (Brusca et al., 2018). Soil nematodes use the soil as its transportation mean to reach and infect the plant root system. The soils where *Rubus* species are cultivated can harbor diverse nematodes such as the *Trychodorus* and *Hemicycliophora* (Múnera, 2000). Most of the damages seem to be caused by the saliva secretion introduced in the tissues of the plants over the nematodes nutrition. They implant saliva inside the cytoplasm, extracting part of the cell content and exhausting photoassimilates (Agrios, 2005). This produces a slight mechanic damage (Volcy, 1997) that causes a decrease in turgor and root biomass, but rarely leads to death. However, the nematodes are usually vectors of a virus that, during absorption of nutrients by the nematodes, can penetrate the plant and develop several pathologies that causes plant death (Munera and Alzate, 2000). Symptoms of nematodes are visible in the root and shoot components of a plant. In the roots component there are tumors, tuber rot, lesions or stains, thickening of the root tip, proliferation of adventious roots, decrease of root biomass and cyst formation. In the shoot component it is possible to observe leaves and stems distortion, foliage stains, galls in seeds leaves and stems, and coloration changes (Munera and Alzate, 2000).

Pratylenchus penetrans

Pratylenchus penetrans; the root-lesion nematode, belongs to genus *Pratylenchus* within the family *Pratylenchidae*. It has a worldwide distribution, but it is especially common in temperate zones. In economic terms, this is the most important nematode that affect the genus Rubus. It has been reported to have killed almost the 25% of a crop population after two years of infection. *P. penetrans* feeds from the plant root causing root lesions that diminishes nutrients and water absorption. It can also cause necrosis and lead to the plant's death. The most efficient mechanism used to control this nematode has been the fumigation of the infected soils, being the crop performance dramatically affected if its proliferation is not controlled (Zasada and Moore, 2014). Despite this is the most effective control mechanism, it can also cause the death of benefical organisms present in soil.

REFERENCES

Abou Ghanem-Sabanadzovic, Nina, Ioannis E. Tzanetakis, and Sead Sabanadzovic. 2013. Rubus Canadensis Virus 1, a Novel Betaflexivirus Identified in Blackberry. *Archives of Virology.* 158 (2): 445–49. https://doi.org/10.1007/s00705-012-1484-7.

Agrios, G. N. 2005. *Plant pathology.* 5th ed. Elsevier Academic Press, New York.

Avilés, M., Castillo, S., Bascon, J., Zea-Bonilla, T., Martín-Sánchez, P. M. and Pérez Jiménez, R. M., 2008. First report of *Macrophomina phaseolina* causing crown and root rot of strawberry in Spain. *Plant Disease.* 57, 382.

Baino, O., Conci, V. and Ramallo, J. 2010. Enfermedades de *Fragaria* x *ananassa* Duch. *Atlas Fitopatológico Argentino.* 3, 739.

Baino, O., Salazar, S., Ramallo, A. and Kirschbaum, D. 2011. First report of *Macrophomina phaseolina* causing strawberry crown and root rot in northwestern Argentina. *Plant Disease.* 95 (11), 1477.

Ban, Yeon-Hee, Mi-Ri Jeon, Ji-Hee Yoon, Jae-Min Park, Hyun-Ju Um, Dae-Soon Kim, Seung-Ki Jung, Young Kim, K, Lee, Jeewon, Min, Jihi and Hoon Kim, Yang. 2008. Differential Proteomic Analysis of Secreted Proteins from Cutinase-Producing *Bacillus* Sp. SB-007. *The Plant Pathology Journal.* 24 (2): 191–201. https://doi.org/10.5423/PPJ.2008.24.2.191.

Bartho J. D., Demitri N., Bellini D., Flachowsky, H., Peil, A., Walsh, M. and Benini, S. 2019. The structure of *Erwinia amylovora* AvrRpt2 provides insight into protein maturation and induced resistance to fire blight by

Malus × robusta 5. *Journal of Structural Biology*. 206 (2):233-242. doi: 10.1016/j.jsb.2019.03.010.

Bascón, J. 2009. Aumento de la incidencia de la podredumbre carbonosa causada por *Macrophomina phaseolina* (Tassi) Goidanich en el cultivo del fresón en la provincia de Huelva. *Agrícola Vergel* 324, 37-40. [Increase in incidence of charcoal rot by *Macrophomina phaseolina* (Tassi) Goidanich in raspberry cultivation of in the Huelva province. *Agrícola Vergel* 324, 37-40].

Benlioğlu, S., Yıldız, A. and Döken, T. 2004. Studies to determine the causal agents of soil-borne fungal diseases of strawberries in Aydin and to control them by soil disinfestation. *Journal of Phytopathology*. 152, 509-513.

Bertaccini, A. and Duduk, B. 2009. Phytoplasma and Phytoplasma Diseases: A Review of Recent Research. *Phytopathologia Mediterranea*. https://doi.org/10.14601/Phytopathol_Mediterr-3300.

Boubourakas, I., Avgelis, A., Kyriakopoulou, P. and Katis, N. 2006. Occurrence of Yellowing Viruses. *Plant Patholog*. 276–83. https://doi.org/10.1111/j.1365 3059.2005.01341.x.

Bowers, G. and Russin, J. 1999 Soybean disease management. En: Heatherly, L.G. y Hodges, H.F. (CRC Press). *Soybean Production in the Midsouth,* 231-271, Washington DC.

Brunt, A., Crabtree, K., Dallwitz, M., Gibbs, A. and Watson, L. 1996. *Viruses of Plants*. CAB International, Wallingford, UK.

Brusca, R., Moore, W. and Shuster, S. 2018. *Invertebrates* (3rd edition). Guanabara Koogan Ltda.

Bühlmann, A., Pothier, J., Rezzonico, F. Smits, T., Andreou, M., Boonham, N., Duffy, B. and Frey, J. 2013. *Erwinia Amylovora* Loop-Mediated Isothermal Amplification (LAMP) Assay for Rapid Pathogen Detection and on-Site Diagnosis of Fire Blight. *Journal of Microbiological Methods*. 92 (3): 332–39. https://doi.org/10.1016/j.mimet.2012.12.017.

Candán, A. 2006. Harvest and post-harvest of cherry fruits. *Fruticultura and Diversificación*. 50: 404 32-38.

Castro, D., Márquez, O., Restrepo, V. and Vélez, G. 1995. Assesment of the phytosanitary status of Andean raspberry (*Rubus glaucus* Benth.) in western Antioquia. *Serie de investigaciones 7*. Fundación de Fomento Agropecuario Buen Pastor. Universidad Católica de Oriente, Rionegro, Antioquia. 15 p.

Cho, C. and Moon, B. 1984. Studies on the wilt of strawberry caused by *Fusarium oxysporum* f. sp. fragariae in Korea. *Korean Journal of Plant Protection.* 23, 74- 81.

Christakopoulos, P., Nerinckx, W., Kekos, D. Macris, B. and Claeyssens, M. 1996. Purification and characterization of two low molecular mass alkaline xylanases from Fusarium oxysporum F3. *Journal of Biotechnology.* 51, 181-189.

Coley-Smith, J., K. Verhoeff and Jarvis, W. 1980. *The biology of Botrytis.* Academic Press. Londres. 318 p.

Crous, 2009. Taxonomy and phylogeny of the genus Mycosphaerella and its anamorphs. *Fungal Diversity.* 38, pp. 1-24.

Davies, D. 2000. The Occurrence of Two Phytoplasmas Associated with Stunted Rubus Species in the UK. *Plant Pathology.* 49 (1): 86–88. https://doi.org/10.1046/j.1365-3059.2000.00418.x.

Deighton, F. 1976. Studies on *Cercospora* and allied genera. VI. *Pseudocercospora* Speg., *Pantospora* Cif. and *Cercoseptoria Petr Mycological Papers.* 140, pp. 1-168.

Den Breeÿen, J., Groenewald, G., Verkley and Crous, P. 2006. Morphological and molecular characterisation of *Mycosphaerellaceae* associated with the invasive weed, Chromolaena odorata. *Fungal Diversity.* 23, pp. 89-110.

Dhingra, O. and Sinclair, J. 1975. Survival of *Macrophomina phaseolina* sclerotia in soil: Effects of soil moisture, carbon: nitrogen ratios, carbon sources, and nitrogen concentrations. *Phytopathology.* 65, 236-240.

Diaz-Lara, A., Martin, R. Rwahnih, M. Vargas, O. and Rebollar-Alviter, Á. 2019. First Evidence of Viruses Infecting Berries in Mexico. *Journal of Plant Pathology.* 1–7. https://doi.org/10.1007/s42161-019-00381-9.

Diaz-Lara, A., Mosier, N., Keller, K. and Martin R. 2015. A Variant of Rubus Yellow Net Virus with Altered Genomic Organization. *Virus Genes.* 50 (1): 104–10. https://doi.org/10.1007/s11262-014-1149-6.

Dong, L., Lemmetty, A., Latvala, S., Samuilova, O and Valkonen, J. 2016. Occurrence and Genetic Diversity of Raspberry Leaf Blotch Virus (RLBV) Infecting Cultivated and Wild Rubus Species in Finland. *Annals of Applied Biology.* 168 (1): 122–32. https://doi.org/10.1111/aab.12247.

Eastwell, K., Villamor, D., McKinney, C. and Druffel, K. 2010. Characterization of an Isolate of Sowbane Mosaic Virus. *Archives of Virology* 155 (12): 2065–67. https://doi.org/10.1007/s00705-010-0820-z.

Ermacora P. and Osler R. 2019. Symptoms of Phytoplasma Diseases. In *Phytoplasmas: Methods and Protocols, Methods in Molecular Biology.* 1:53–67. New York: Humana Press Inc. https://doi.org/10.1007/978-1-4939-8837-2_1.

Fang, X., Phillips, D., Li, H., Sivasithamparam, K. and Barbetti, M. 2011. Comparisons of virulence of pathogens associated with crown and root diseases of strawberry in Western Australia with special reference to the effect of temperature. *Scientia Horticulturae.* 131, 39-48. doi: 10.1016/j.scienta.2011.09.025.

Farnsworth, D., Hamby, K., Bolda, M., Goodhue, R., Williams, J. and Zalom. F. 2017. "Economic Analysis of Revenue Losses and Control Costs Associated with the Spotted Wing Drosophila, Drosophila suzukii (Matsumura), inthe California raspberry industry. *Pest Manag Sci.* 73 (6):1083-1090. doi: 10.1002/ps.4497.

Franco, G. and Giraldo, M. 2000. *El cultivo de la mora*. Corporación Colombiana de Investigación Agropecuaria (Corpoica), Manizales. pp. 53, 43-55. *[Growing raspberries.* Corporación Colombiana de Investigación Agropecuaria (Corpoica), Manizales. pp. 53, 43-55].

Fravel, D., Olivain, C. and Alabouvette, C. 2003. *Fusarium oxysporum* and its biocontrol. *New Phytologist* 157, 493–502.

Garibaldi, A., Gilardi, G. and Gullino, M. 2004. Seed transmission of Fusarium oxysporum f. sp. lactucae. *Phytoparasitica.* 32, 61-65.

Gergerich, R. 2007. Introducction to Vegetables Virus, The Invisible Foe. *The Plant Health Instructor.* https://doi.org/10.1094/phi-i-2008-0122-01.

Gobernación de Antioquia. 2014. *Guideline for growing raspberry under Good Agricultural Practices.* Secretaría de Agricultura y Desarrollo Rural, Medellín. 113p.

Golzar, H., Phillips, D. and Mack, S. 2007. Occurrence of strawberry root and crown rot in western Australia. *Australasian Plant Disease Notes.* 2 (1), 145-147.

González, F., Walls, S. and Mancilla, M. 2005. Detecction by PCR-RFLP of *Fusarium oxysporum* f. sp. *fragariae*, the causal agent of fusariosis in raspberry. *Boletín Micológico.* 20, 63-72.

González, G., Concha, C., Valenzuela, M., Cordero, L. Pico, J., Caceres, P. and Garcia, R. 2017. Distribution and Frequency of Tomato Ringspot Virus (ToRSV) in Different Varieties of *Rubus idaeus* in the Maule Region, Chile. *Revista de La Facultad de Ciencias Agrarias.* 49 (X): 143–46.

Gordon, T. and Martyn, R. 1997. The evolutionary biology of Fusarium oxysporum. *Annual Review of Phytopathology.* 35, 111-128.

Graham, J., and Brennan, R. 2018. *Raspberry.* Edited by Julie Graham and ex Brennan. Cham: Springer International Publishing. https://doi.org/10.1007/978-3-319-99031-6.

Guo, M., Bian, X. Wu, X. and Wu, M. 2011. Agrobacterium-Mediated Genetic Transformation: History and Progress. Edited by Mara Alvarez. *Genetic Transformation.* InTech. https://doi.org/10.5772/868.

Guo, Minliang, Jingyang Ye, Dawei Gao, Nan Xu, and Jing Yang. 2019. Agrobacterium-Mediated Horizontal Gene Transfer: Mechanism, Biotechnological Application, Potential Risk and Forestalling Strategy. *Biotechnology Advances.* Elsevier Inc. https://doi.org/10.1016/j.biotechadv.2018.12.008.

Harper, S., Delmiglio, C., Ward, L. and Clover, G. 2011. Detection of Tomato Black Ring Virus by Real-Time One-Step RT-PCR. *Journal of Virological Methods.* 171 (1): 190–94. https://doi.org/10.1016/j.jviromet.2010.10.023.

Hawksworth, D., Crous, P., Redhead, S., Reynolds, D., Samson, R. et al. 2011. The Amsterdam declaration on fungal nomenclature. *IMA Fungus.* pp. 105-112.

Hincapié, O., Saldarriaga, A. and Díaz, C. 2017. Biological, botanical and chemical alternatives for the control of blackberry (*Rubus glaucus* Benth.) diseases. *Revista Facultad Nacional de Agronomía Medellín.* 70:8169-8176.

Hukkanen, A., Pietikäinen, L., Kärenlampi S and Kokko, H. 2006. Quantification of downy mildew (Peronospora sparsa) in Rubus species using real-time PCR. *European Journal of Plant Pathology.* 116:225-235

Hurtado, Nathalia Cardona, Álvarez, Gloria Edith Guerrero, & Gutiérrez, Ana María López. (2019). Identificación de Peronospora sparsa y evaluación del contenido de fenoles en frutos de mora de castilla afectados por este microorganismo. *Revista Ceres.* 66(1), 11-17. https://dx.doi.org/10.1590/0034-737x201966010002. [Identification of *Peronospora sparsa* and evaluation of pehnolic content in infected Andean raspberry fruits. *Revista Ceres.* 66(1), 11-17. https://dx.doi.org/10.1590/0034-737x201966010002].

Ica, 2011. *Phytosanitary management of cultivated Andean raspberry (Rubus glaucus Benth): Indications for the winter season.* Bogotá, Produmedios. 32p.

James, D, Howell, W. and Gi, M. 2001. Molecular Evidence of the Relationship between a Virus Associated with Flat Apple Disease and Cherry Rasp Leaf Virus as Determined by RT-PCR, *Plant Disease.* 85 (1):47-52. doi: 10.1094/PDIS.2001.85.1.47.

Jarvis, W. 1977. Botryotinia and Botrytis Species: taxonomy, physiology, and pathogenicity. Monograph 15. *Research Branco.* Canada Department of Agriculture. 195 p.

Jennings, D. 1988. *Raspberries and blackberries. Their breeding, diseases and growth.* Academic Press, London. 230 p.

Jiménez-Gasco, M., Milgroom, M. andJiménez-Díaz, R. 2004. Stepwise evolution of races in Fusarium oxysporum f. sp. ciceris inferred from fingerprinting with repetitive DNA sequences. *Phytopathology.* 94, 228–235.

Jones, A. and Roberts, I. M. 1976. Ultrastructural Changes and Small Bacilliform Particles Associated with Infection by Rubus Yellow Net

Virus. *Annals of Applied Biology*. 84 (3): 305–10. https://doi.org/10.1111/j.1744-7348.1976.tb01773.x.

Jones, A., McGavin, W., Gepp, V., Zimmerman, M. and Scott, S. 2006. Purification and Properties of Blackberry Chlorotic Ringspot, a New Virus Species in Subgroup 1 of the Genus Ilarvirus Found Naturally Infecting Blackberry in the UK. *Annals of Applied Biology*. 149 (2): 125–35. https://doi.org/10.1111/j.1744-7348.2006.00078.x.

Kalischuk, M., Fusaro, A., Waterhouse, P., Pappu, H. and Kawchuk, L. 2013. Complete Genomic Sequence of a Rubus Yellow Net Virus Isolate and Detection of Genome-Wide Pararetrovirus-Derived Small RNAs. *Virus Research*. 178 (2): 306–13. https://doi.org/10.1016/j.virusres.2013.09.026.

Kämpfer, P. and Glaeser, S. 2013. Prokaryote Characterization and Identification. In *The Prokaryotes: Prokaryotic Biology and Symbiotic Associations,* 123–47. Springer-Verlag Berlin Heidelberg. https://doi.org/10.1007/978-3-642-30194-0_6.

Katan, J. 1971. Symptomless carriers of the tomato Fusarium wilt pathogen. *Phytopathology*. 61, 1213-1217.

Koike, S. 2008. Crown rot of strawberry caused by Macrophomina phaseolina in California. *Plant Disease*. 92(8), 1253-1253. doi: 10.1094/pdis-92-8-1253b.

Kondo, T., Yamashita, K. and Sugiyama, S. 2011. First Report of Impatiens Necrotic Spot Virus Infecting Chrysanthemum (Chrysanthemum Morifolium) in Japan. *Journal of General Plant Pathology*. 77 (4): 263–65. https://doi.org/10.1007/s10327-011-0317-y.

Kuzmanović, N., Puławska, J. Prokić, A. Ivanović, M., Zlatković, N., Jones, J. and Obradović, A. 2015. "Agrobacterium Arsenijevicii Sp. Nov., Isolated from Crown Gall Tumors on Raspberry and Cherry Plum." *Systematic and Applied Microbiology* 38 (6): 373–78. https://doi.org/10.1016/j.syapm.2015.06.001.

Kuzmanović, N., Puławska, J. Smalla, K. and Nesme, X. 2018. Agrobacterium Rosae Sp. Nov., Isolated from Galls on Different Agricultural Crops. *Systematic and Applied Microbiology*. 41 (3): 191–97. https://doi.org/10.1016/j.syapm.2018.01.004.

Lemanceau, P., Bakker, P., DeKogel, W., Alabouvette, C. and Schippers, B. 1993. Antagonistic effect of nonpathogenic Fusarium oxysporum Fo47 and pseudobactin 358 upon pathogen Fusarium oxysporum f. sp. dianthi. *Applied Enviromental Microbiology*. 59, 74-82.

Leslie, J. and Summerell, B. 2006. *The Fusarium laboratory manual,* 1st edn. Backwell Publishing, Iowa, USA.

Linck, H. and Reineke, A. 2019. Rubus Stunt: A Review of an Important Phytoplasma Disease in Rubus Spp. *Journal of Plant Diseases and Protection*. July. https://doi.org/10.1007/s41348-019-00247-3.

Linck, H., Krüger, E. and Reineke, A. 2016. Detection and Management of Rubus Stunt, a Phytoplasma Disease in Rubus Species. *Acta Horticulturae*. 1133 (May): 511–14. https://doi.org/10.17660/Acta Hortic.2016.1133.80.

Mann, R., Blom, J., Bühlmann, A., Plummer, K., Beer, S., Luck, J., Goesmann, A., Frey, J., Rodoni, B., Duffy, B. and Smits, T. 2012. Comparative Analysis of the Hrp Pathogenicity Island of Rubus- and Spiraeoideae-Infecting *Erwinia Amylovora* Strains Identifies the IT Region as a Remnant of an integrative conjugative element. *Gene* 504 (1): 6–12. https://doi.org/10.1016/j.gene.2012.05.002.

Martin, R., MacFarlane, S., Sabanadzovic, S., Quito, D., Poudel, B. and Tzanetakis. I. 2013. Viruses and Virus Diseases of Rubus. *Plant Disease*. 97 (2): 168–82. https://doi.org/10.1094/PDIS-04-12-0362-FE.

Martins, P., Merfa, M., Takita, M. and De Souza, A. 2018. Persistence in Phytopathogenic Bacteria: Do We Know Enough? *Frontiers in Microbiology. Frontiers Media S.A.* https://doi.org/10.3389/fmicb. 2018.01099.

Mäurer, R. and Seemüller, E. 1995. Nature and Genetic Relatedness of the Mycoplasma-like Organism Causing Rubus Stunt in Europe. *Plant Pathology*. 44 (2): 244–49. https://doi.org/10.1111/j.1365-3059.1995. tb02775.x.

Mcgavin, W., Cock, P. and Macfarlane. S. A. 2011. Partial Sequence and RT-PCR Diagnostic Test for the Plant Rhabdovirus Raspberry Vein Chlorosis Virus. *Plant Pathology*. 60 (3): 462–67. https://doi.org/10. 1111/j.1365-3059.2010.02387.x.

McNicol, R. J.; B. Williamsom and A. Dolan. (1985). Infection of red raspberry styles and carpels by Botrytis cinerea and its possible role in post-harvest grey mould. *Annals of Applied Biology.* 106, 49-53.

Medina, C., Matus, J. Zúñiga, M, San-Martín, C. and Arce-Johnson, P. 2006. "Occurrence and Distribution of Viruses in Commercial Plantings of Rubus, Ribes and Vaccinium Species in Chile." *Ciencia e Investigación Agraria.* 33 (1): 19–24. https://doi.org/10.7764/rcia.v33i1.324.

Mena, A., Palacios de Garcia, M. and Gonzalez, M. 1975. Strawberry root diseases caused by Fusarium oxysporum f. sp. fragariae and Rhizoctonia fragariae. *Revista Agronómica del Noroeste Argentino.* 12, 299-307.

Mertely, J., Seijo, T. and Peres, N. 2005 First report of Macrophomina phaseolina causing a crown rot of strawberry in Florida. *Plant Disease.* 84, 434.

Montoya, C., Hincapié, L. and Uribe, V. 2003. *Main pest and diseases in cultivated raspberries.* Bogotá, Instituto Colombiano Agropecuario. 20p.

Munera, G. and Alzate, Á. 2000. Nematodes associated to moderate cold climate fruits. *CDTF.* 136–40. http://hdl.handle.net/20.500.12324/16778.

Murant, A., Jones, A., Martelli, G. and Stace-Smith, R. 1996. Nepoviruses: General Properties, Diseases, and Virus Identification. In *The Plant Viruses.* 99–137. Boston, MA: Springer US. https://doi.org/10.1007/978-1-4899-1772-0_5.

Nam, M., Kang, Y., Lee1, I., Kim, H. and Chun, Ch. 2011. Infection of daughter plants by Fusarium oxysporum f. sp. fragariae through runner propagation of strawberry. *Korean Journal of Horticultural Science and Technolog.* 29 (3), 273-277.

Nelson, K. 1956. The effect of de Botrytis infection on the tissue of Tokay grapes. *Phytophatology.* 46, 223- 229.

Nita Lazar M., Heyraud, A., Gey, C., Braccini, I. and Lienart, Y. 2004. Novel oligosaccharides isolated from Fusarium oxysporum L. rapidly induce PAL activity in Rubus cells. *Acta Biochimica Polonica.* 51:625-634.

Oyarzun, P., Postma, J., Luttikholt, A. and Hoogland, A. 1994. Biological control of foot and root rot in pea caused by Fusarim solani with nonpathogenic Fusarium oxysporum isolates. *Canadian Journal of Botany.* 72, 843-852.

Pastrana, A. 2014. Incidencia y epidemiología de nuevos hongos patógenos de fresa en la provincia de Huelva. *Desarrollo de herramientas biotecnológicas y aplicación de otras estrategias de control. Doctorado.* Instituto de Formación Agraria y Pesquera de Andalucía La Torres – Tomejil. Sevilla, pp. 18-21. [Incidence and epidemiology of new pathogen raspberry fungi in the province of Huelva. *Development of biotechnological tools and application of control strategies. Ph D Thesis.* Instituto de Formación Agraria y Pesquera de Andalucía La Torres – Tomejil. Sevilla, pp. 18-21].

Pearson, R. and Goheen, A. 1996. *Plagas y enfermedades de la vid.* Madrid, España. Editorial Mundi Prensa. Pág. 91. [*Grapevine Pests and Diseases.* Madrid, España. Editorial Mundi Prensa. Pág. 91].

Pratt, R. 2006. A direct observation technique for evaluating sclerotium germination by Macrophomina phaseolina and effects of biocontrol materials on survival of sclerotia in soil. *Mycopathologia.* 162, 121-131

Quinatoa, N. 2015. *Assessment of botrytis control (Botrytis cinerea) in cultivated Andean raspberry (Rubus glaucus Benth) by Trichoderma yemas in Misquillí de la parroquia Santa Rosa, province of Tungurahua.* Universidad Técnica de Ambato. Facultad de Ciencias Agropecuarias, Ambato Ecuador. 83 p.

Quito-Avila, D. Lightle, D. and Martin, R. 2014. Effect of Raspberry bushy dwarf virus, Raspberry leaf mottle virus, and Raspberry latent virus on Plant Growth and Fruit Crumbliness in 'Meeker' Red Raspberry, *Plant Dis.* 98(2):176-183. doi: 10.1094/PDIS-05-13-0562-RE.

Quito-Avila, D. and Martin, R. 2012. Real-Time RT-PCR for Detection of Raspberry Bushy Dwarf Virus, Raspberry Leaf Mottle Virus and Characterizing Synergistic Interactions in Mixed Infections. *Journal of Virological Methods.* 179 (1): 38–44. https://doi.org/10.1016/j.jviromet.2011.09.016.

Ramos, B., Algar, E., Gutiérrez, F. Bonilla, A., Lucas, J. and García, D. 2015. Bacterial bioeffectors delay postharvest fungal growth and modify total phenolics, flavonoids and anthocyanins in blackberries. *Food Science and Technology*. 06:437-443.

Ramos, B., Garcia, A., Gutierrez F., Lucas J., Bonilla, A. and Garcia, D. 2014. Annual changes in bioactive contents and production in field-grown blackberry after inoculation with Pseudomonas fluorescens. *Plant Physiology and Biochemistry*. 74:01-08.

Roberto, L., Reveles-Torres, R., Velásquez-Valle, J., and Mauricio-Castillo, A. 2014. *Fitoplasmas: Otros Agentes Fitopatógenos*. 41 p. [*Phytoplasmas: Other Phytopathogenic Agents*. 41 p].

Rodríguez K., Silva, H., Boyzo, J., Segura S., Leyva S. and Rebollar, A. 2017. Molecular detection of Peronospora sparsa in sources of primary inoculum and components of resistance in wild blackberry species. *European Journal of Plant Pathology*. 149: 845-851.

Rodríguez, A., Perestelo, F., Carnicero, A., Regalado, V., Pérez, R., De la Fuente, G. and Falcon, M. 1996. Degradation of natural lignins and lignocellulosic substrates by soil-inhabiting fungi imperfecti. *FEMS Microbiology and Ecology*. 21, 213-219.

Sabanadzovic, S. and Abou Ghanem-Sabanadzovic, N. 2009. Identification and Molecular Characterization of a Marafivirus in Rubus Spp. *Archives of Virology*. 154 (11): 1729–35. https://doi.org/10.1007/s00705-009-0510-x.

Saldarriaga, A. and Bernal, J. 2000. Diseases associated to cultivated Andean raspberry (*Rubus glaucus* Benth.) in the Departamento de Antioquia. p. 132-135. In: *Proceedings of the Third Symposium on Moderate Cold Climate Fruits*. CDTF, Manizales.

Sanchez, S., Gambardella, M., Henriquez, J. and Diaz, I. 2013. First Report of Crown Rot of Strawberry Caused by Macrophomina phaseolina in Chile. *Plant Disease*. 97(7), 996-996. doi: 10.1094/pdis-12-12-1121-pdn.

Schulz, M. and Freitas Chim, J. 2019. Nutritional and Bioactive Value of Rubus Berries. *Food Bioscience*. 31: 100438. https://doi.org/10.1016/j.fbio.2019.100438.

Sharifi, K. and Mahdavi, M. 2011. First report of strawberry crown and root rot caused by Macrophomina phaseolina in Iran. *Iranian Journal of Plant Pathology,* 47 (4), 161.

Smith, M. 1948. Plant Virus Diseases. *Nature.* 161 (4102): 945. https://doi.org/10.1038/161945a0.

Sochor, M., Trávníček, B. and Manning, J. 2018. Biosystematic Revision of the Native and Naturalised Species of Rubus L. (Rosaceae) in the Cape Floristic Region, South Africa. *South African Journal of Botany.* 118: 241–59. https://doi.org/10.1016/j.sajb.2018.07.015.

Strik, B., Chad, E. Finn, E., Clark, J. and Bañados, M. 2008. "Worldwide Production of Blackberries." *Acta Horticulturae.* 777: 209–17. https://doi.org/10.17660/actahortic.2008.777.31.

Susaimuthu, J., Tzanetakis, I., Gergerich, R. and Martin, R. 2008. *"A Member of a New Genus in the Potyviridae Infects Rubus"* 131: 145–51. https://doi.org/10.1016/j.virusres.2007.09.001.

Swiss Institute of Bioinformatics. *Viral Zone.* Retrieved October 24, 2019, from https://viralzone.expasy.org.

Tamayo, P. 2001. Main diseases in tomato tree, Andean raspberry and lulo tree in Colombia. Medellín, Centro de Investigación La Selva. 40p. (*Technical Bulletin 12*).

Tamayo, P. 2003. Main diseases in tomato tree, Andean raspberry and lulo tree in Colombia. *Technical Bulletin Corpoica 20.* Rionegro, Antioquia, pp. 16-25.

Tamayo, P. 2009. [Diseases. In: *Memoirs of the Technology Update Seminar: Cultivation, Agroindustry and comercialization of Andean raspberry.* Corpoica, Rionegro, Antioquia, Colombia. CD.

Tamayo, P. and Peláez, A. 2000. Characterization of damages and lossess caused by fruit diseases on Andean raspberry (*Rubus glaucus* Benth.) in Antioquia. pp. 174-178. In: *Proceedings of the Third Symposium on Moderate Cold Climate Fruits.* CDTF, Manizales.

Tidona, C., and Gholamreza, D. 2011. The Springer Index of Viruses. *The Springer Index of Viruses.* https://doi.org/10.1007/978-0-387-95919-1_53.

Trigiano, R., Windhim, M. and Windhim, A. 2007. Plant Pathology Concepts and Laboratory Exercises. *Plant Pathology Concepts and Laboratory Exercises.* Second edi. CRC Press. https://doi.org/10.1201/b15333.

Undurraga, P. and Vargas, S. 2013. *Strawberry Grow Guide*. Boletín INIA N° 264. 108 p. Instituto de Investigaciones Agropecuarias INIA, Centro Regional de Investigación Quilamapu, Chillán, Chile.

Volcy, C. 1997. Nematodes. Volume 1. *The ABC of Nematology.* First edition. Ecográficas Ltda., Medellín.

Ward, J., Boone, W., Moore, P. and Weber, C. 2012. "Developing Molecular Markers for Marker Assisted Selection for Resistance to Raspberry Bushy Dwarf Virus (RBDV) in Red Raspberry." *Acta Horticulturae.* 946: 61–66. https://doi.org/10.17660/ActaHortic.2012.946.6.

Weintraub, P. and Jones, P. 2010. *Phytoplasmas: Genomes, Plant Hosts and Vectors.* Wallingford, UK: CABI.

Williamsom, B., McNicol R. and Dolan, A. 1987. The effect of inoculating flowers and developing fruits with Botrytis cinerea on post-harvest grey mould of red raspberry. *Annals of Applied Biology.* 11, 285-294.

Williamson, M., Fernandez-Ortuño, D. and Schnabel, G. 2012. First Report of Fusarium Wilt of Strawberry Caused by *Fusarium oxysporum* in South Carolina. *Plant Disease.* 96(6), 911-911. doi: 10.1094/pdis-02-12-0164-pdn.

Wingfield, M., DeBeer, Z., Slippers, B., Wingfield, B., Groenewald, J., Lombard, L. and Crous, P. 2012. One fungus, one name promotes progressive plant pathology. *Molecular Plant Pathology.* doi: 10.1111/J.1364-3703.2011.00768.X.

Winks, B. and Williams, Y. 1965. A wilt of strawberry caused by a new form of *Fusarium oxysporum*. *Queensland journal of agricultural and animal sciences* 22, 475-479.

Wintermantel, W. 2004. Emergence of Greenhouse Whitefly (*Trialeurodes vaporariorum*) Transmitted Criniviruses as Threats to Vegetable and Fruit Production in North America. *APSnet Feature Articles*, no. June. https://doi.org/10.1094/apsnetfeature-2004-0604.

Wyllie, T. 1988. Charcoal rot of Soybean current status. In Wyllie, T. D. y Scott, D. H. (APS Press). *Soybean Diseases of the North Central Region*, 106–113. St. Paul, MN.

Zasada, I., and Moore, P. 2014. “Host Status of Rubus Species and Hybrids for the Root Lesion Nematode, PratylenchusPenetrans.” *Hort Science*. 49 (9): 1128–31. https://doi.org/10.21273/hortsci.49.9.1128.

Zhao, X., Zhen, W., Qi, Y., Liu, X., and Yin, B. 2009. Coordinated effects of root autotoxic substances and Fusarium oxysporum Schl. f. sp. fragariae on the growth and replant disease of strawberry. *Frontiers of Agriculture in China* 3, 34-39.

Zveibel, A., Mor, N., Gnaymen, N. and Freeman, S. 2012. Survival, host-pathogen interaction, and management of *Macrophomina phaseolina* on strawberry in Israel. *Plant Disease*. 96 (2), 265-272.

In: Rubus: An Overview
Editor: Davet Chabot

ISBN: 978-1-53617-376-5

Chapter 4

IMPROVEMENT IN *RUBUS* THROUGH MODERN TOOLS

Samriti Sharma[*]
Department of Biotechnology, Chandigarh group of colleges,
Landran, Mohali, India

ABSTRACT

Rubus is the main member of Rosaceae family having basic chromosome number seven. It is native to temperate and subtropical regions of eastern Asia. *Rubus* is a medicinally important wild fruit crop used in the treatment of fever, cough, sore throat, epithelial cancer and cardiovascular disease. Important traits in *Rubus* are genetically encoded i.e., Yield, insect-pest resistance and quality linked parameters. But breeding in *Rubus* is tedious and time consuming due to highly heterozygous nature of this perennial fruit crop. Recent advances in the development of DNA or molecular markers have irreversibly changed the disciplines of breeding in *Rubus*. Increasing the role of genomics and genome editing technologies in *Rubus* presents new opportunities to develop more focused molecular tools for gene discovery and deployment which is not possible with conventional breeding techniques. So, the

[*] Corresponding Author's Email: 1992samritisharma@gmail.com.

advancement in genomic information and modifications in linked properties directly improve crop by improving various traits in *Rubus*.

Keywords: genetic diversity, genome editing, genetic transformation, *in-vitro* cultures, *Rubus*

INTRODUCTION

Rubus is derived from Latin word Ruber, which means Red. *Rubus* is a member of Rosaceae family which also includes apples, roses, pears, cherries, peaches, plums and strawberries. *Rubus* is one of the most diverse genera which include approximately 800 species, native to temperate and subtropical regions of eastern Asia. (Marulanda et al. 2012). The genus *Rubus* is divided into 12 subgenera having basic chromosome number seven, however polyploidy also exists. The polyploidy level varies from diploid in subgenera *Idaeobatus, Dalibarda,* and *Anoplobatus* to tetradecaploid (98) in *Dalibardastrum, Malachobatus* and *orobatus* (Thompson 1995; Thompson 1997) etc. These species are characterized as diploid and polyploid on the basis of its chromosome numbers, chromosomes distribution and pollens formation (Longley, 1924).

Among the *Rubus* genus, only a few species have been domesticated and utilized in breeding programs those are raspberries, blackberries, dewberries, artic fruits and flowering raspberry (Jenning, 1988). Most of these plants have woody stems which prickle like roses, spines, bristles and gland tipped hairs (Jennings, 1983). The fruits of *Rubus* are known as brambles because it forms an aggregate of drupelets. It is a medicinally important fruit crop due to antioxidant, antimicrobial, anti-inflammatry, antidiabetic and antiaging properties (Ferlemi et al. 2016; Grochowski et al. 2016; Cho et al. 2016; Chen et al. 2017). It provides protection against a variety of diseases such as lung diseases, cardiovascular and neurological diseases. The high content of bioactive compounds (ellagitannins, phenolic acids, flavonoids, vitamin C& E and carotenoids) in fruits is responsible for protective effects (Probst, 2015).

The potential health benefits of berry fruit consumption is now the subject of considerable public investment by various non-governmental organizations and governmental sectors (Verma et al. 2014). However for plant breeders, it is essential to produce new cultivars or improve previous cultivar with increased desirable components and agronomic traits (Yield, abiotic and biotic stress resistance). This becomes possible only become possible if information related to number of genes involved, function of genes, location of genes and diversity within genes is known (Stinchcombe et al. 2007).Molecular markers are now widely used to track loci and genome regions in several crop-breeding programmes, as molecular markers tightly linked with a large number of agronomic and disease resistance traits. Various types of molecular markers have been used in *Rubus* to study variations present in individuals or group of individuals or populations by a specific method or a combination of methods (Samriti et al. 2017). Several genetic Linkage maps composed of different types of molecular markers are available for raspberry (Ward et al. 2013; Woodhead et al. 2008; Bushakra et al. 2015), and one is available for blackberry (Castro et al. 2013). But, crop specific molecular markers in *Rubus* are available in limited number. So, the crop specific markers for *Rubus* can be designed by means of sequencing or transcriptomics. Another genomic tool which can also be used in crop improvement programmes of *Rubus* is CRISPRs. CRISPR is extremely simple, economical, and versatile method for specific genome editing processes.

In-Vitro Characterization of *Rubus*

Traditionally propagation techniques by suckering and cutting can be limited by inability of genotypes to root, transmittance of disease or pest to daughter plants or being too slow and unreliable for rapid multiplication or bulk up new cultivars. In-vitro propagation of *Rubus* is commonly used in commerce (Table 1). It is a best method to provide new cultivars more quickly into the marketplace, and to introduce clean planting material for more conventional subsequent propagation.

Table 1. Details of various varieties of *Rubus*

Sr. No	Name of cultivars	Discription
Cultivated varieties		
1	Boysen-berry	The major identifier of the berry fruits is its characteristic aroma. Fruits early December to mid January
2	Chester thornless	Thornless variety were derived from Merton thornless Flavour of berry is acidic until full ripe Fruits from mid January to mid march
3	Dirksen thornless	Dirksen thornless was derived from Merton thornless Duration of Fruiting from early January to late February. Flavour is acidic until fruit ripe and having poor aroma
4	Karaka black	Having good disease resistance against botrytis. Mild acidic flavour of under-ripe berry
5	Loch ness	Harvesting of Fruits in late December to late January. The flavour is distinct and aromatic as compare to hybrid berries.
6	Logan-berry	Susceptible for downy mildew, crown gall and anthracnose. Having long harvesting period from mid November to mid January. Better suited for processing and having acidic taste until it is over ripe.
7	Marion-berry	Fruits are very much sensitive to UV light. Harvesting period is from late December to mid January.
8	Silvan black-berry	Excellent acidic/aromatic flavour and is nearest flavour to wild weed blackberry. Fruit is soft and easily damaged at full-ripe. Harvesting of fruits during early December to midJanuary
Minor commercial varieties		
1	Ranui	Derived from cross between Aurora×Marion This fruit is not widely grown because it is not fit into the raspberry and blackberry category due to its appearance. Flavour is sweet/acidic The fruits were harvested in December.
2	Tayberry	Harvesting of fruits in late November and end December. Excellent flavour and aroma. Flavour is also outstanding when fruit is processed.
Superseded varieties		
1	Black satin	Genetically thornless variety derived from Merton thornless. Fruit is harvested from late December to mid February. Flavour and aroma of fruit is poor i.e., more acidic and less sugar content.
2	Lawton-berry	Harvested from early January to mid February. The size of fruit is smaller than other berries but having greater shelf life. Flavour of berry is acidic/aromatic which is close to wild blackberry flavour.
3	Murrin-dindi	Fruits are harvested from mid of December to end of January. Fruits have greater shelf life. Fruits having acidic/aromatic flavour but having poor intensity of aroma.
4	Smooth-stem	Harvesting of fruits from early February to late march. Flavour is acidic/aromatic but undistinctive.
5	Thorn-free	Harvesting of fruit from mid January to mid march Flavour of berry is acidic but lack of aroma and distinctiveness.
6	Waldo	Fruits were harvested in January. Suitable for home garden. Leaves of plant not fruit are prone to mite and thrips damage.

In-Vitro Cultures of Raspberry

In-vitro propagation in raspberry was first conducted by Anderson (1980). Earlier, virus free varieties have been created by breeding, heat therapy which was still used clean order germplasm, and micropropagation has become an efficient and widely used method for propagating newly developed selections (Wang et al. 2008). The susceptibility of Raspberry to various viral diseases makes it necessary to use tissue culture for elite and certified plant production (Table 2). Despite of various publications on *Rubus* culturing, there are still difficulties that influence *in-vitro* propagation of some *Rubus* cultivars (McPheeters et al. 1988; Reed, 1990; Gonzalez et al. 2000; Tian et al. 2005; Zawadzka and Orlikowska, 2006).

Table 2. Number of diseases in *Rubus*

Sr No	Disease	Microorganism	Infection	Symptoms
Virus disease				
1	Raspberry Bushy Dwarf disease	Raspberry Bushy Dwarf virus	Transmitted by pollen and spread by bees and other insects	Yellow flecks or splashes on leaves Poor drupelet set causing malformed fruits
Fungal disease of fruit				
2	Grey mould	*Botrytis cinerea*	Infection begins during flowering and remains latent in withered floral parts	Grey fungal growth frequently affecting only a few druplets, and then more commonly at the aperture end of the fruit at the "collar"
3	Fruit Anthracnose	*Elsinoe veneta*	Exacerbated by rain during ripening and different from common grey mould by the absence of hyphae.	Hard grey patch on the top of druplets around the point where the style meets the fleshy skin of the drupelet
Fungal disease of canes and foliage				
4	Cane Anthracnose	*Elsinoe veneta*	Preventative spraying should commence during winter, when cane are dormant	Blotchy, purple and grey lesions on the primocanes of raspberry, Loganberries and blackberry.
5	Raspberry yellow rust	*Phragmidium rubi-idaei*	Conditions is exacerbated by cold wet weather	It appears firstly as yellow-orange pustules on the upper side of raspberry leaves in spring.
6	Blackberry rusts	*Phragmidium violaceum*		Symptoms begin with yellow-red blotches on the upper darker to red purple spots with yellow or brown centre
7	Septoria leaf spot	*Septoria rubi*	Damage is more severe following prolonged wt periods and control begins by encouraging foliage to dry quickly after rain	Evenly shaped lesions on blackberry canes and leaves, usually with a whitish centre and purple-brown margin

Table 2. (Continued)

Sr No	Disease	Microorganism	Infection	Symptoms
Root diseases				
8	Cane and leaf rust	*Kuehneola uredinis*	Kuehneola can be distinguish from phragmidium rust because absence of prominent red brown spot on the upper surface of affected leaves	Producing yellow or orange spore bodies on stems and the undersides of leaves.
10	Phytophthora root rot	*Phytophthora cryptogea*	Start with fine, white feeder roots which form in late winter-spring.	Loss of functional roots Floricanes die after bud burst Leaves browning and chlorosis
11	White Root rot	*Vararia*	Disease is most severe in warm dry soils	Masses of white hyphae Affect plants wilt and die
12	Arimalia	Arimalia species	Very difficult to control Eradication relies on removable of dead and drying plants.	Dead and rotting wood Fungal growth is characterized of white or yellowish mycelia sheets on the main roots and crown.
Pests				
13	Two spotted mite	Tetranychus urticae	Found in warm weather	Pale spot to appear on the upper side of leaves Reduce growth and quality of bud
14	Dried fruit bulte	*Carpophilus spp*	Pest of drying and rotting fruit	Rotting of fruit No insecticide is register tillnow
15	Lopper caterpillars	*Epiphyas postvittana*	Their prevalence varies considerably from one season to another	Their main damage to raspberry is cosmetic
16	Earwings	Forticula auricularia		Chew into druplets leaving the druplets hollowed out with the seeds half exposed.
17	Green vegetable bettle	*Nezara viridula*	Attacks raspberry fruit by sucking of sap, causing drupelets to shrivel	Control by clean Farm management No chemicals are register till now
18	Plague thrips	*Thrips imaginis*	They migrate by wind	Cause damage to by sucking sap from all flower parts
19	Dock sawfly	*Ametastegia glabrata*	Remove dock weeds which act as a primary host.	Bores holesin primocanes late in the growing season, causing growth above the entry oint to die. Control by farm hygiene
20	Rutherglen bug	*Nysius vinitar*	Strongly influenced by weather, being most prevalent in dry spring weather following wet winter	Suck sap from fruit and leaves.
21	Fruit tree Borer	Maroga Melanostigma	Control should start with either removing or treating their major hosts in the area such as plums or black wattle tree	Boring into mature raspberry primocane slims, causing snap during training and fruiting

Plant materials used for propagation are firstly tested for virus existence. During spring and summer, Bud segments from primocane growth are collected. Bud collected after differentiation are fail to produce vegetative growth because at this time, they exhibit reproductive organs and are no longer suitable for initiation. For winter initiation, vegetative buds collected are suitable for vegetative growth initiation.

Plant material collected from field are mainly contaminated, so sterilization has been made by immersion in 70% ethylalcohol for 5 seconds and agitation in a concentration of 1% sodium hypochlorite and two drops of tween 20 for 20min (Wu et al. 2009) or by rinsing hypochlorite and agitation twice for ten minute in 1-5% sodium hypochlorite depending on the sensitivity of explants. The sterilized explants were placed into culture media after splicing under aseptic conditions. The cultures are then kept in a growth room having temperature 26+_1°C with 16+_8h of photoperiod. When the desired multiplication has been achieved plants with 3-5 leaves can transfer to rooting media under aseptic conditions. Roots may be form as soon as 11 days after the plantlets are transferred. These plantlets remain rooting media for 4 weeks. After then they are transferred to the greenhouse.

The main problem with raspberry culturing is mutation which is frequently responsible for crumbled fruiting and worthless for commercial cultivation. To overcome from this problem, care must be taken not to multiply adventitious buds formed around the base of the explants and cultures should be sub-cultured in a limited number.

In Vitro Cultures of Blackberry

Micropropagation of Blackberry usually done with sequential shoot tip proliferation because it is most efficient method for rapid proliferation and avoid somaclonal variation. Broome and Zimmerman (1978) evaluated 'Smoothstem', 'SIUS 64-39-2 selection', 'Black Satin' and 'Dirksen Thornless' for micropropagation variables and found variation in explant proliferation for different media. *In-vitro* regeneration and proliferation for broad number of *Rubus* accessions (256) was evaluated by Reed (1990) and reported that 69% of blackberry genotypes proliferated successfully on

Murashige and Skoog (1962) basal salts. Tissue culture of meristem culture and axillary bud can also be done but it is suffered from various problems. Shoot tip culture helps us to overcome these problems which makes it more popular for propagation. Meristem culture was mainly done in conjugation with Thermotherapy to produce virus free propagation stocks.

Future research has been done to address other *in vitro* improvements and issues such as improved efficiency (Radmann et al. 2003), avoidance of somaclonal variation (McPheeters et al., 1988), evaluation of various explants material (Harper, 1978), or somatic embryogenesis (Cantoni et al. 1993).

In vitro culturing through embryo and seed has been used to increased survival rate of hybrid seeds (Clark et al. 2007). First, successfully cultures embryos of blackberry were obtained between cross *R. allegheniensis*× *R. idaeus* (Selonen and tigerstedt, 1989). Galletta et al. 1986 achieved good germination using embryo culture with upto 50% of transferable seedlings from crossing. The USDA-ARS in Oregon used *in vitro* methods for germination which involves cutting the top of the seeds off to physically remove some of the barriers to germination.

Limited research has been done n regeneration system of blackberry which was used in genetic engineering. When leaves were incubated in a thidiazuron (TDZ) pre-treatment medium for 21 days (before culture on regeneration medium), regeneration efficiency of explants were increased up to 70%. Place this cultured explants into growth room having 23°C temperature for 28 days and 16+_8h of photoperiod.Recent work has been done to produce a transformed callus and chimeric plantlets from several week of TDZ and darkness period treatment before co-cultivation. Unfortunately, they are yet to regenerated fully transformed plants. We are not aware any transgenic that has been released till date and there are no active breeding efforts employing transgenic methods in blackberry improvement.

Genetic Transformation in *Rubus*

Recent advances in biotechnology have allowed incorporating useful genes into raspberry genome (Kempler et al. 2011). Genetic transformation is the process of altering genetic composition of the plant material creating transgenic plants. This process has been made in raspberry by introducing RBDV resistance gene into cultivars 'Meeker'. Fruit set in genetic transformed 'Meeker' remains a problem, however, some plants appeared commercially viable, but they have been not yet released on commercial scale (Kempler et al. 2011).

Genetic transformation had been used to increased nutraceutical values of number of food crops like incorporation of genes for beta carotene, Zinc, Iron and Lysine into rice. Several vitamins and micronutrients have been identified in raspberry having positive health impacts. Research has been showed that Raspberry contained significant amount of antioxidant (anthocyanins and polyphenols) such as ellagic acid which provides anti-aging and anti-cancer benefits (Kempler et al. 2011). It has been suggested that the genetic transformation technique could be used to increase the concentration of nutraceuticals in raspberry (Hall et al. 2009). However, the biggest challenges to commercially development and release of transgenic varieties are public caution toward consuming genetically transformed food crops (Kempler et al. 2011).

DNA Meta-Barcoding for *Rubus*

The medicinal plants are used to cure various ailments in traditional system but nowadays herbal industry suffering from adulteration and substitution of closely related species with medicinal plants. If the adulteration occurred, it directly decreases the efficiency of drugs sometimes it may be lethal, if it is substituted with toxic adulterants. Hence for effective outcomes correct formulation is important for the medicinal plants. There are number of traditional methods for medicinal plants identifications such as organoleptic methods (identification by taste, smell, sight or touch), macroscopic and microscopic methods (e.g., texture, colour and shape) and chemical profiling (HPLC-UV, HPLC-MS and TLC). However, neither of the method gave validate results for identification of the related species in processed product because for macroscopic and microscopic methods trained personal is required whereas chemical profile maybe affected by physiological and storage conditions. In this case authentication is obtained at DNA level as compare to RNA level, because DNA is a stable macromolecule which is not affected by external factors and found in all tissue. Therefore development of DNA based markers provides authentication in medicinal plants identifications.The short DNA sequences either from nuclear or organelles genome for identification of medicinal plants is known as DNA barcoding. Paul Hebert of University of Guelph in 2003 first proposed the term 'DNA barcode' for taxon identifiers.

For meta-barcoding in *Rubus*, genomic DNA was isolated using Cetyltrimethylammonium bromide (CTAB) method from green fresh and healthy leaves. Amplicons were prepared using g and h universal primers from P6 loop region of TmL (UAA) chloroplast intron. The illumina adaptors were ligated to the purified amplicons for second round of PCR and all amplicons were purified using MiniElute R kit. After purification illumina 2500 was performed with purified product. The raw data than quality filtered with PRINSEQ version 0.20.2. The PANDASEQ version 2.7 used for overlapping pair end reads detection. The noise ratio is removed with USEARCH for taxonomic assignments and search against reference database with MegaBLAST (Figure 1). For taxonomic group identification a minimum of 10 sequences had to be assigned across all samples to be included in analysis. The overall differences in breadth or detection biases were identified using ANOVA tests blocks by sampling instances and post hoc turkey's tests. Jaccard dissimilarity used for calculations of compositional agreement among DNA markers at order, family and gene levels using "Betadisper" in the vegan package. The dissimilarity matrices were also developed with PCoAs for spatial medians identification (measure the distance of each point to the median).

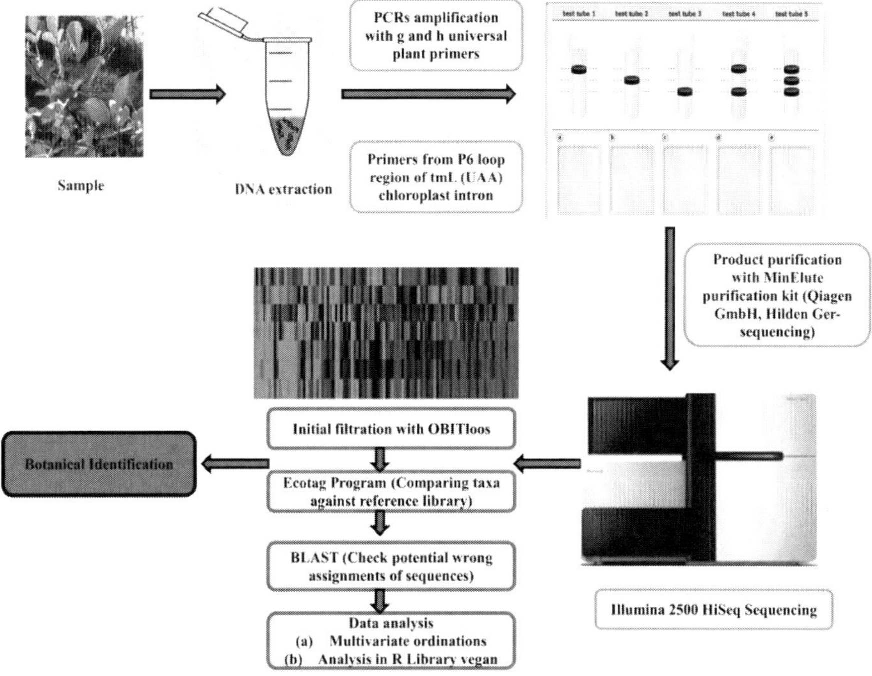

Figure 1. DNA metabarcoding method for taxa classification.

HIGH-THROUGHPUT TRANSCRIPTOME ANALYSIS IN *RUBUS*

RNA-seq is considered a powerful molecular tool for investigating non-model species that have little information available for genetic studies (Mutz et al. 2013). The RNA sequencing (RNA-seq) and whole-transcriptome sequencing using next-generation sequencing (NGS) technologies has started to reveal the complex dynamics and landscape of the transcriptome with unprecedented level of accuracy and sensitivity (Martin et al. 2010). Thetranscription and translation is the two-step process, by which the information in genes flows into proteins: DNA → RNA → protein.The identification of candidate genes related to agronomic traits and their transcriptional profile might reveal new hypotheses about genetic mechanisms that control proteins and metabolites biosynthesis. Currently, high-throughput mRNA sequencing techniques (RNA-seq) have been widely used in studies of plant transcriptomes (Seco et al. 2015). In the data generation phase, total RNAs or mRNAs are fragmented and converted into a library of cDNAs containing sequencing adaptors (Levin et al. 2010). The millions to billions of short reads from one end or both ends of the cDNA fragments are produced by sequencing of cDNA library by next-generation sequencers (Figure 2).

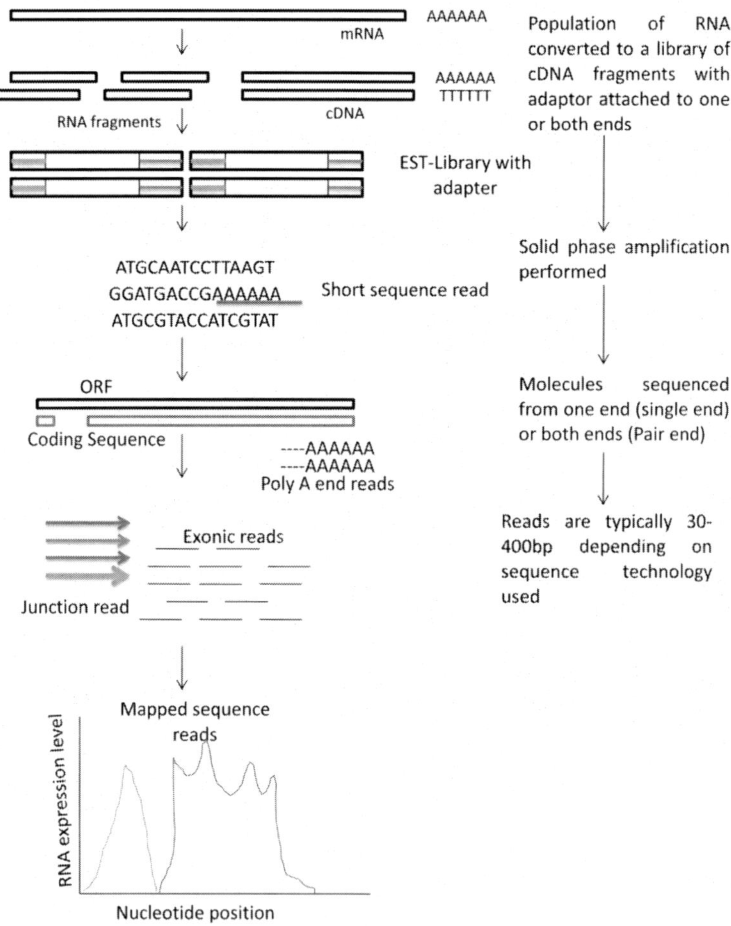

Figure 2. Overview of sequencing.

In the data analysis phase, these short reads are pre-processed to remove sequencing errors and other artifacts. The reads are subsequently assembled to reconstruct the original RNAs and to assess their abundance ('expression counting') and its role in cellular pathway (Figure 3) Large numbers of molecular markers and sequence data from across the genome are playing an important role in population genomic studies to find out fine genetic variations and the genetic basis of traits (Whitlock, 2014). Nevertheless, whole genome sequencing is still largely impractical for most eukaryotes and there is lack of genomic resources for most non-model organisms. The genomic level data for non-model organisms can be efficiently provided by transcriptome sequencing or Expressed Sequence Tag (EST) or those with genomic characteristics prohibited to whole genome sequencing. The EST sequencing is an alternative approach for whole genome sequencing because EST sequences are derived from exonic region whereas most of the eukaryotic genome is non-codingthat render analysis and interpretation of data more difficult (Bouck et al. 2007).

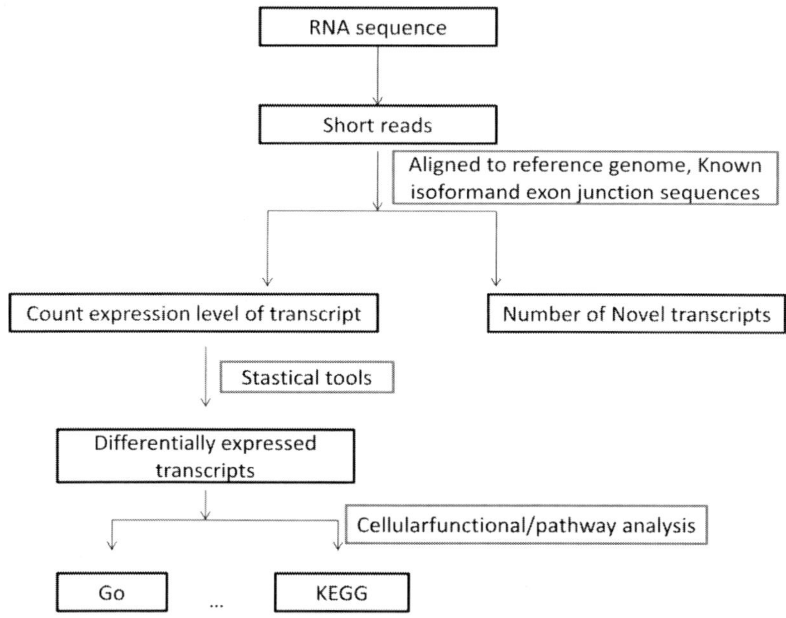

Figure 3. RNA sequence analysis mechanism.

ESTs thus have high functional information content, and often correspond to genes with known or predicted functions (Bouck et al. 2007, Sharma et al. 2019). Large collections of EST sequences have proven invaluable for gene annotation and discovery (Bouck et al. 2007, Emrich et al. 2007), comparative genomics (Vera et al. 2008), development of molecular markers (Sharma et al. 2018, Sharma et al. 2008), and for population genomic studies of genetic variation associated with adaptive traits (Singh et al. 2019).

Nonetheless, until recently, traditional approaches involved the utilization of cloning, cDNA library construction and Sanger sequencing runs for the development of ESTs which are highly costly and time consuming (Bouck et al. 2007). The first blackberry EST library from cDNA library of 18,432 clones generated from the cultivar 'Merton Thornless', a progenitor of many thornless commercial cultivars were developed by Lewers *et al.* 2008. They designed 673 primer pairs, out of which 33 primer pairs were tested with two blackberry cultivars. They utilized these markers to construct the genetic map which could be used for marker-assisted breeding in blackberry. Bushakra *et al.* 2015 were utilized Leaf and meristem tissue from cultivar 'Heritage' of red raspberry (*Rubus idaeus*) and cultivar of 'Bristol' black raspberry (*R. occidentalis*) for the development of EST-SSRs. The 1149 and 2358 cDNA sequences of red raspberry and black raspberry were used to generate 131 and 288 primer pairs, respectively. The sequence analysis revealed that the SSR-containing genes have different functions and share maximum sequence identity with strawberry genes as compare with other Rosaceae species. Sharma et al, 2019 developed EST-SSRs markers for *Rubus ellipticus* of 'Kumarhatti' collection. A total of 304 primer pairs were generated out of which 68 randomly selected primers were used for polymorphism and transferability studies. The *Rubus* specific gene derived markers will promote construction of linkage maps (transferable markers) and helps in the manipulation of agricultural important traits in this economical important genus.

GENOME EDITING IN *RUBUS*

Genome editing comprises predicted changes in the gene sequence or precise insertion of exogenous DNA with the goal of inactivating gene(s), generating functional alleles, replacing mutant alleles or site-specific transgene integration. Genome editing is more precise method of crop improvement than conventional crop breeding methods or standard genetic engineering (transgenic or GM). Genome-editing technologies includes chimeric DNA/RNAgene repair oligonucleotides (Gamper et al. 2000), zinc-finger nucleases (Urnov et al. 2010), homingendo nucleases (Hafez et al. 2012), transcription activator-like effector nucleases often referred to as TALENs (Bogdanove et al. 2011), and clustered regularly inter-spaced short palindromic repeats (CRISPR)/Cas9 (Shan et al. 2013).

These editing technologies exhibit a number of advantages over previous technologies such as RNA interference (RNAi) and RNA antisense which were designed to simply reduce gene transcript abundance. Current gene editing technologies can not only be designed to reduce transcript abundance, but to completely remove or even replace entire genes within a genome. Such advancements provide researchers with new avenues to decipher gene function and tailor traits for crop improvement. For many plant scientists CRISPR/Cas9 has become the gene editing technique of choice due to its simplicity and high efficiency (Hsu et al. 2013, Belhaj et al. 2013, Bortesi et al. 2015).

CRISPR stands for clustered regularly interspaced short palindromic repeats which is associated with Cas9 nuclease. CRISPR/Cas system is divided into three types: I, II, III but type II system is commonly required to degrade DNA that matches a single guide RNA (SgRNA). Double or single short RNA molecule directs Cas9 nuclease to recognize desirable target DNA site for genome modification and transcriptional control.gRNA spacer sequence of 20 nucleotide paired with target DNA strand by 5'-end leading sequence following PAM site i.e., NGG. Cas9 endonuclease has two domains; one is HNH nuclease domain for cleavage in target strand and other is RuvC like nuclease domain to create double stranded break at target locus. Initially this mechanism was discovered in Bacteria to protect invasion pathogen. This nicked DNA fragment is repaired by host cell machinery by non-homologous end joining (NHEJ) and homology direct repair pathway to create gene knockout or introduce a specific genetic modification through homologous recombination with a DNA donor.

The CRISPR/Cas9 system is very useful to get regulatory committee approval because it works in such a manner that it does not leave any exogenous DNA in the plants which is the major problem with genetic engineering. *Rubus* is a suffered from variety of diseases (Table 2) which directly effects its production like powdery mildew*Sphaerotheca macularis*(Fr.) Jaczewski in the red raspberry (Keep, 1988), yellow rust (Anthonyet al. 1986), root rot in red raspberry (Barritt et al. 1979; Duncan et al. 1987), disease caused by *Leptosphaeria coniothyrium*in red raspberry (Jennings et al. 1989), cane spot (*Elsinoe veneta*) in red raspberry (Jennings et al. 1988) and cane and foliar diseases in red Raspberry (Williamson et al. 1992).

Nowaday's new virus line known as blackberry yellow vein diseasehas been detected in blackberry which is responsible for leaf mottling, chlorotic ringspots and curved midribs (Hassan et al. 2017). The characteristic and phylogenetic analysis revealed its close relationship to recognized members of the genus Emaravirus. The detection protocols for BYVD were developed for its presence in plants collected from USA.So, resistant from these diseases can be made through artificially synthesized gRNAs which can mutate, activate, or repress almost any genomic locus. Number of repeated sequences of CRISPRs locus is separated by spacer sequence from previous exposure of virus or phage. Future infection with same pathogen will responsible for transcription of this spacer sequence into full length CrRNA. This full length CrRNA is processed into specific small RNA molecules which find its complementary pathogen DNA. Complementary base pair annealing activate Cas9 endonuclease to cleave the targeted DNA sequence. This is known a plant own defense mechanism. These all properties make the CRISPR/Cas9 the ideal genome engineering system for large-scale forward genetic screening.

CONCLUSION

Use of genomics and genome editing in horticulture and agriculture crops will help to identify desirable traits or modified already existing traits leading to healthier and more productive system. Designing of efficient and effective strategies provides greater integration between molecular genetics and plant breeding. These tools provide better understanding of the genetics of numerous agronomic important traits like pest and disease resistance, flavonol production affecting fruit quality, the timing of flowering and development of molecular markers linked to desirable traits. Molecular genetics and genomics studies in *Rubus* have benefited the whole production system, from breeding to nurseries and growers, by providing tools to skilful variety identification. So, new genomic principles play an important role in modification and discovery of number of genes in plant genome in order to increases the quality and quantity of agricultural production. Finally, we believe that the identification of genes and genes responsible for various characters will have a strong impact in future *Rubus* breeding programmes.

REFERENCES

Andersen, J. R. and Lubberstedt, T. (2003). Functional markers in plants. *Trends in Plant Science,* 8(11): 554-560.

Anthony, V. M., Williamson, B., Jennings, D. L., Shattock, R. C. (1986).Inheritance of resistance to yellow rust (*Phragmidium rubi-idaei*) in red raspberry. *Annals of Applied Biology,* 109:365-374.

Barbazuk, W. B., Emrich, S. J., Chen, H. D., Li, L. and Schnable, P. S. (2007). SNP discovery via 454 transcriptome sequencing. *Plant Journal,* 51(5): 910-918.

Barritt, B. H., Crandall, P. C. and Bristow, P. R. (1979). Breeding for root rot resistance in red raspberry. *Journal of the American Society for Horticultural Science*, 104: 92 - 94.

Belhaj, K., Chaparro-Garcia, A., Kamoun, S. and Nekrasov, V. (2013). Plant genome editing made easy: targeted mutagenesis in model and crop plants using the CRISPR/Cas system. *Plant Methods,* 9:1-10.

Bogdanove, A. J. and Voytas, D. F. (2011). TAL effectors: customizable proteins for DNA targeting. *Science,* 333:1843–6.

Bortesi, L. and Fischerk, R. (2015). The CRISPR/Cas9 system for plant genome editing and beyond. *Biotechnology Advances*, 33:41–52.

Bouck, A. and Vision, T. (2007). The molecular ecologist's guide to expressed sequence tags. *Molecular Ecology,* 16(5): 907-924.

Bushakra, J. M., Lewers, K. S., Staton, M. E., Zhebentyayeva, T. and Saski, C.A. (2015). Developing expressed sequence tag libraries and the discovery of simple sequence repeat markers for two species of raspberry (*Rubus* L.). *BMC Plant Biology*, 15: 1-11.

Castro, P., Stafne, E.T., Clark, J.R. and Lewers, K. S. (2013). Genetic map of the primocane-fruiting and thornless traits of tetraploid blackberry. *Theor Appl Genet,* 126: 2521–2532.

Chen, X.,Wu, X.,Ouyang, W.,Gu, M.,Gao, Z.,Song, M.,Chen, Y.,Lin, Y.,Cao, Y. andXiao, H. (2017). Novel ent-Kaurane Diterpenoid from*Rubus*corchorifolius L. f. Inhibits Human Colon Cancer Cell Growth via Inducing Cell Cycle Arrest and Apoptosis. *J Agric Food Chem,* 65(8):1566-1573. doi: 10.1021/acs.jafc.6b05376. Epub 2017 Feb 13.

Cho, S. H., Hong, J. H., Noh, Y. W., Lee, E., Lee, C. S. and Lim, Y. T. (2016). Raspberry-like poly(γ-glutamic acid) hydrogel particles for pH-dependent cell membrane passage and controlled cytosolic delivery of antitumor drugs. *Int J Nanomedicine,* 11:5621-5632.

Duncan, J. M., Kennedy, D. M. and Seemuller, E. (1987). Identities and pathogenicities of Phytophthora spp. causing root rot of red raspberry. *Plant Pathology,* 36: 276 - 289.

Emrich, S. J., Barbazuk, W. B., Li, L. and Schnable, P. S. (2007). Gene discovery and annotation using LCM-454 transcriptome sequencing. *Genome Research,* 17: 69-73.

Ferlemi, A. V.and Lamari, F. N. (2016). Berry Leaves: An Alternative Source of Bioactive Natural Products of Nutritional and Medicinal Value. *Antioxidants,* 5(2): E17. doi: 10.3390/antiox5020017.

Gamper, H. B., Parekh, H.,Rice, M. C.,Bruner, M.,Youkey, H. andKmiec, E. B. (2000).The DNA strand of chimeric RNA/DNA oligonucleotides can direct gene repair/conversion activity in mammalian and plant cell-free extracts. *Nucleic acids research*, 28(21): 4332–4339.

Grochowski, D. M., Paduch, R., Wiater, A., Dudek, A., Pleszczyńska, M., Tomczykowa, M., Granica, S., Polak, P. and Tomczyk, M. (2016). *In Vitro* Antiproliferative and Antioxidant Effects of Extracts from *Rubus*caesius Leaves and Their Quality Evaluation. *Evid Based Complement Alternat Med.* doi: 10.1155/2016/5698685.

Hafez, M. and Hausner, G. (2012). Homing endonucleases: DNA scissors on a mission. *Genome,* 55(8): 553-569.

Hall, H. K., Hummerm K. E., Jamiesonm A. R., Jenning, S. N. and Weber, C. A. (2009). Raspberry Breeding. In: Janick J. (Ed) *Plant Breeding Reviews*, 32: 339-353.

Hassan, M., Bello, P. L. D., Keller KE, Martin RR, Sabanadzovic S, Tzanetakis IE (2017) A new, widespread emaravirus discovered in blackberry. *Virus Res.* 235: 1-5. doi: 10.1016/j.virusres.2017.04.006.

Hsu, D. P., Lander, E. S. and Zhang, F. (2014). Development and Applications of CRISPR-Cas9 for Genome Engineering. *Cell,* 157(6): 1262–1278.

Jennings, D. L. and Brydon, E. (1989).Further studies on breeding for resistance to *Leptosphaeria coniothyrium*in red raspberry and related species. *Annals of Applied Biology,* 115: 499-506.

Jennings, D. L. and Ingram, R. (1983). Hybrids of *Rubus parviflorus*(Nutt.) with raspberry and Blackberry, and the inheritance of spinelessness derived from this species. *Crop Research (Hort Res.),* 23:95-101.

Jennings, D. L. and McGregor, G. R. (1988).Resistance to cane spot (*Elsinoe veneta*) in red raspberry and its relationship to resistance to yellow rust (*Phragmidium rubi-idaei*). *Euphytica,* 37:173-180.

Keep, E. (1988). Primocane (autumn)-fruiting raspberries: A review with particular reference to progress in breeding. *Journal of Horticultural Science,* 63:1-18.

Kempler, C., Hall, H. K. and Finn, C. E. (2011). Raspberries. In: Badenes ML, Byme DH (Eds) *Handbook of Plant Breeding* (Vol 8), Springer, New York, in press.

Levin, J. Z.,Yassour, M., Adiconis, X., Nusbaum, C., Thompson, D. A., Friedman, N., Gnirke, A. and Regev, A. (2010). Comprehensive comparative analysis of strand-specific RNA sequencing methods. *Nature Methods*, 7: 709–715.

Lewers, K. M., Saski, C. A., Cuthbertson, B. J., Henry, D. C., Staton, M. E., Main, D. S., Dhanaraj, A. L., Rowland, L. J. and Tomkins, J. P. (2008). A blackberry (*Rubus* L.) expressed sequence tag library for the development of simple sequence repeat markers. *BMC Plant Biology,* 8: 69-77.

Longley, A. E. (1924). Cytological studies in the Genus *Rubus*. *American journal of botany,* 11: 249-282.

Martin, J. A. and Wang, Z. (2011). Next-generation transcriptome assembly. *Genetics,* 12: 671-682.

Marulanda, M., Lopez, A. M. and Uribe, M. (2012). Molecular characterization of the Andean blackberry, *Rubus glaucus*, using SSR markers. *Genetic and Molecular Research,* 11:322-331.

Mutz, K. O., Heilkenbrinker, A., Lonne, M., Walter, J. G. and Stahl, F. (2013). Transcriptome analysis using next-generation sequencing. *Current Opinion in Biotechnology,* 24: 22-30.

Namroud, M. C., Beaulieu, J., Juge, N., Laroche, J. and Bousquet, J. (2008). Scanning the genome for gene single nucleotide polymorphisms involved in adaptive population differentiation in white spruce. *Molecular Ecology,* 17(16):3599-3613.

Novaes, E., Drost, D. R., Farmerie, W. G., Pappas, G. J. J., Grattapaglia, D., Sederoff, R. R. and Kirst, M. (2008). High-throughput gene and SNP discovery in Eucalyptus grandis, an uncharacterized genome. *BMC Genomics,* 9:312.

Probst, Y. (2015). A review of the nutrient composition of selected *Rubus* berries. *Nutrition and Food Science,* 45 (2): 242-254.

Samriti., Kaur, R., Shilpa., Malhotra, E.V., Poonam., Thakur, D. and Kumar, K. (2017). Assessment of Genetic Diversity in *Rubusellipticus* (Smith) Using Molecular Markers. *Proc Indian Natn Sci Acad,* 83: 1-11.

Sharma, S., Dobhal, S. and Thakur, S. (2018). Analysis of genetic diversity in parents and hybrids of *Populus deltoides* Bartr. Using microsatellite markers. *Applied Biological Research,* 20(3): 262-270.

Sharma, S. and Sharma, A. (2018). Molecular markers based plant breeding. *Advances in Research,* 16(1): 1-15.

Sharma, S., Kaur, R., Solanke, A. K. U., Dubey, H., Tiwari, S. and Kumar, K. (2019). Transcriptome sequencing of Himalayan Raspberry (*Rubus ellipticus*) and development of simple sequence repeat markers. *3 Biotech,* 9(4): doi: 10.1007/s13205-019-1685-9.

Singh, T. J., Gupta, T. and Sharma, S. (2019). Development and purity identification of hybrids by using molecular marker in wild pomegranate (*Punica granatum* L.). *Scientia Horticulturae,* 247: 436–448.

Seco, D. G., Zhang, Y., Manero, G. F. J., Martin, C. and Solano, B. R. (2015). RNA-Seq analysis and transcriptome assembly for blackberry (*Rubus* sp. Var. Lochness) fruit. *BMC Genomics,* 16: 1-11.

Shan, Q.,Wang, Y., Li, J., Zhang, Y.,Chen, K.,Liang, Z., Zhang, K.,Liu, J.,Xi, J. J.,Qiu, J. L. and Gao, C. (2013). Targeted genome modification of crop plants using a CRISPR-Cas system. *Nature Biotechnology,* 31: 686–688

Stinchcombe, J. R. and Hoekstra, H. E. (2007). Combining population genomics and quantitative genetics: finding the genes underlying ecologically important traits. *Heredity,* 100:158-170.

Thompson, M. M.(1995). Chromosome numbers of *Rubus*species at the National Clonal Germplasm Repository. *HortScience,* 30:1447–1452.

Thompson, M. M.(1997). Survey of chromosome numbers in *Rubus* (Rosaceae: Rosoideae). *Annals of the Missouri Botanical Garden* 84:128–164.

Urnov, F. D., Rebar, E. J., Holmes, M. C., Zhang, H. S. and Gregory, P. D. (2010). Genome editing with engineered zinc finger nucleases. *Nat Rev Genet,* 11:636–46.

Vera, J. C., Wheat, C. W., Fescemyer, H. W., Frilander, M. J., Crawford, D. L., Hanski, I. and Marden, J. H. (2008). Rapid transcriptome characterization for a nonmodel organism using 454 pyrosequencing. *Molecular Ecology,* 17(7): 1636-1647.

Verma, R., Gangrade, T., Punasiya, R. and Ghulaxe, C. (2014). *Rubus fruticosus*(blackberry) use as an herbal medicine. *Pharmacogn Rev.* 8(16): 101–104.

Ward, J. A., Bhangoo, J., Fernandez, F. F., Moore, P., Swanson, J. D., Viola, R., Velasco, R., Bassil, N., Weber, C. A. and Sargent D. (2013). Saturated linkage map construction in *Rubus* idaeus using genotyping by sequencing and genome-independent imputation. *BMC Genomics,* 14: 1-14.

Whitlock, R. (2014). Relationships between adaptive and neutral genetic diversity and ecological structure and functioning: a meta-analysis. *Journal of Ecology,* 102: 857–872.

Williamson, B., Jennings, D. L. (1992).Resistance to cane and foliar diseases in red Raspberry (*Rubus idaeus*) and related species. *Euphytica.* 63:59-70.

Woodhead, M., McCallum, S., Smith, K., Cardle, L., Mazzitelli, L. and Graham, J. (2008). Identification, characterisation and mapping of simple sequence repeat (SSR) markers from raspberry root and bud ESTs. *Mol Breed,* 22: 555-563.

INDEX

A

B

C

D

E

F

G

H

I

L

M

N

O

P

R

S

T

U

V

W

Y

Z

Related Nova Publications

Germination: Types, Process and Effects

Editors: Rosalva Mora-Escobedo, PhD, Cristina Martinez, and Rosalía Reynoso

Series: Plant Science Research and Practices

Book Description: *Germination: Types, Process and Effects* is a book that brings together the contribution of new and relevant information from many experts in the fields of food and biological sciences, nutrition, and food engineering, to provide the reader with the latest information of fundamental and applied research in the role of edible seeds and discuss the benefits of consuming them.

Hardcover ISBN: 978-1-53615-973-8
Retail Price: $230

Phytochemicals: Plant Sources and Potential Health Benefits

Editor: Iman Ryan

Series: Plant Science Research and Practices

Book Description: The opening chapter of *Phytochemicals: Plant Sources and Potential Health Benefits* discusses macronutrients and micronutrients from plants along with their benefits to human health.

Hardcover ISBN: 978-1-53615-478-8
Retail Price: $230

To see a complete list of Nova publications, please visit our website at www.novapublishers.com

Related Nova Publications

Plant Dormancy: Mechanisms, Causes and Effects

Editor: Renato V. Botelho

Series: Plant Science Research and Practices

Book Description: Dormancy is a mechanism found in several plant species developed through evolution, which allows plants to survive in adverse conditions and ensure their perpetuation.

Hardcover ISBN: 978-1-53615-380-4
Retail Price: $160

Micropropagation: Methods and Effects

Editor: Valdir M. Stefenon, Ph.D.

Series: Plant Science Research and Practices

Book Description: Plant micropropagation is one of the most classical and widespread biotechnological tools used around the world. Undoubtedly, this technique brought quite important advances to our knowledge about morphological, physiological and developmental patterns of plants, to the progress of genetic breeding and to the establishment of the genetic engineering, among others.

Softcover ISBN: 978-1-53614-968-5
Retail Price: $82

To see a complete list of Nova publications, please visit our website at www.novapublishers.com